1

Nelson Maths for **NEW ZEALAND**

Connecting Number Facts to Applications

Student Book 1

Lynne Petersen

Peter Hughes

Judy Duncan

Name: _____

NELSON
CENGAGE Learning

Australia • Brazil • Japan • Korea • Mexico • Singapore • Spain • United Kingdom • United States

I look at the pictures and know the number.
I do not need to count.

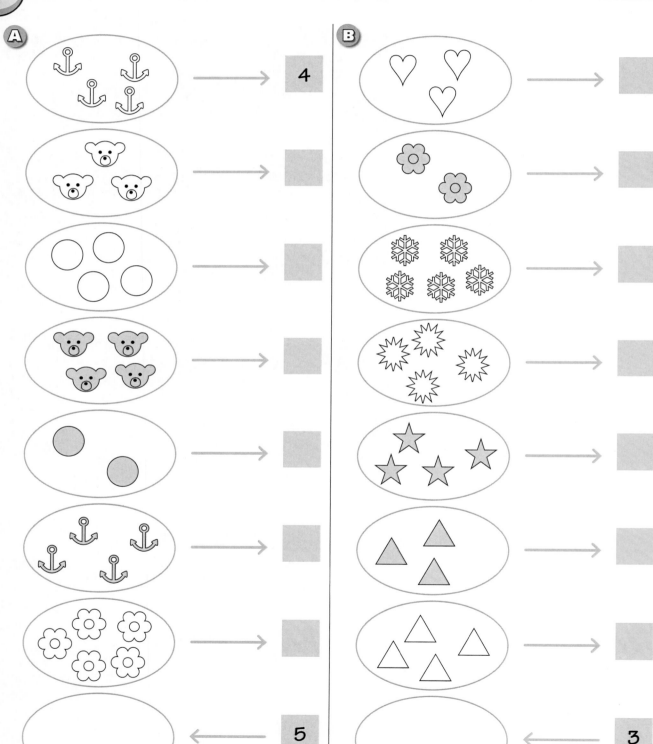

Adding with numbers up to five

I use pictures to add
small numbers together.

 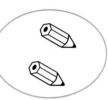   3 + 2 =

    2 + 2 =

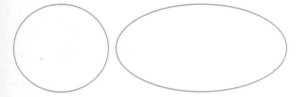 1 + 3 = 4

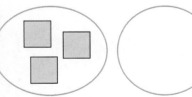   2 + 3 =

  3 + ☐ = 4

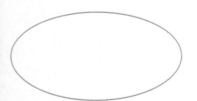   ☐ + 2 = 5

  ☐ + 4 = 5

Adding using five

I use pictures with five to add.
I do not need to count.

  ⟶ 5 + 3 = ☐

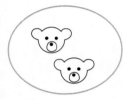  ⟶ 2 + 5 = ☐

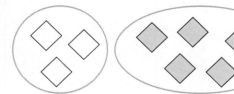 ⟶ ☐ + 5 = 8

 ⟵ 1 + 5 = 6

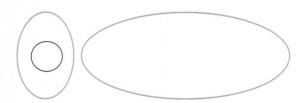 ⟶ 3 + ☐ = ☐

 ⟵ 5 + 2 = 7

 ⟵ 5 + 4 = 9

Adding one more

I think about the numbers over five.
I add on one more.

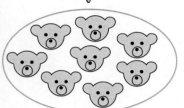 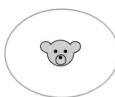 $\longrightarrow$ $8 + 1 = \boxed{}$

  $\longrightarrow$ $6 + \boxed{} = \boxed{}$

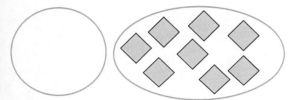 $\longleftrightarrow$ $\boxed{} + 8 = 9$

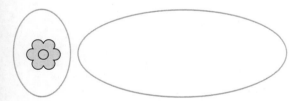 $\longleftrightarrow$ $1 + \boxed{} = 6$

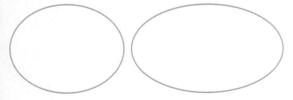 $\longleftrightarrow$ $\boxed{} + 6 = 7$

 $\longleftrightarrow$ $5 + \boxed{} = 6$

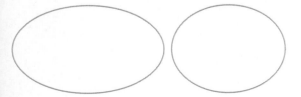 $\longleftrightarrow$ $7 + \boxed{} = 8$

Unit 1

I look at a ten frame. I know the number by seeing circles and empty spaces.

Unit 1

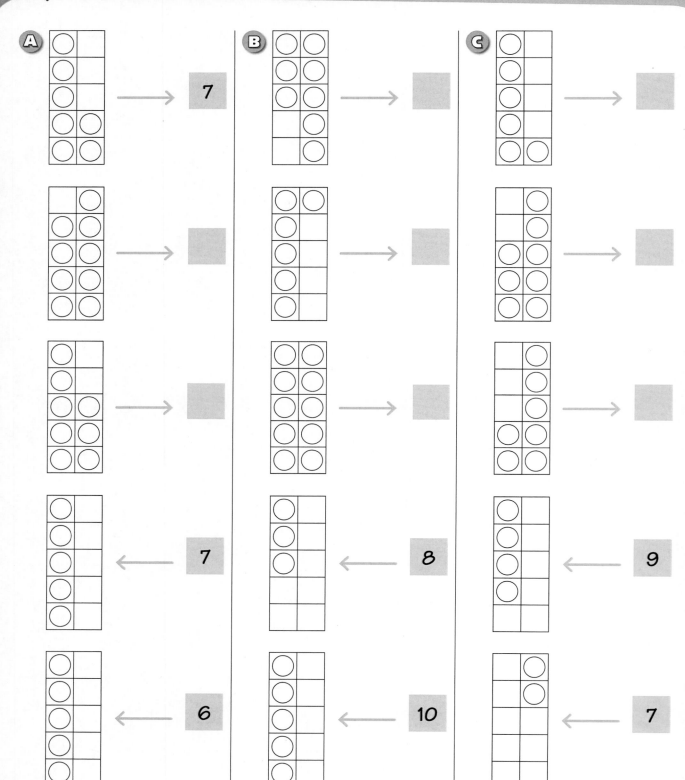

Adding with ten frames

I use ten frames to add.

A

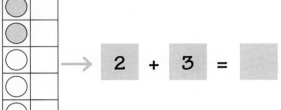

$2 + 3 = \boxed{}$

$6 + 1 = \boxed{}$

$1 + \boxed{} = 6$

$2 + \boxed{} = 7$

$2 + 3 = \boxed{}$

B

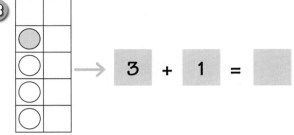

$3 + 1 = \boxed{}$

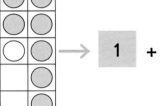

$1 + 7 = \boxed{}$

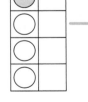

$3 + \boxed{} = 5$

$5 + \boxed{} = 10$

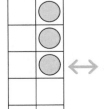

$3 + 4 = \boxed{}$

Unit 1

To take away three, I colour three circles with a dark colour. I think about the circles that are left.

Unit 1

$$5 - 3 = 2$$

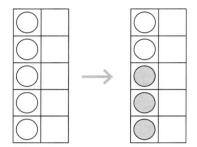

A

$$5 - 2 = \boxed{}$$

B

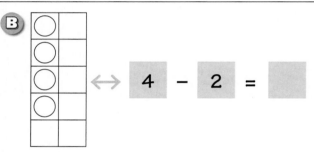

$$4 - 2 = \boxed{}$$

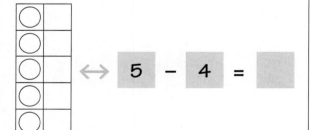

$$5 - 4 = \boxed{}$$

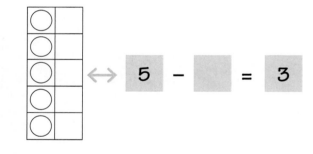

$$5 - \boxed{} = 3$$

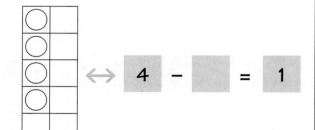

$$4 - \boxed{} = 1$$

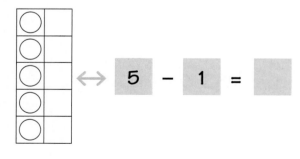

$$5 - 1 = \boxed{}$$

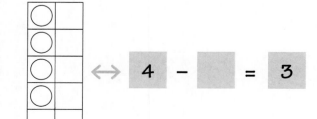

$$4 - \boxed{} = 3$$

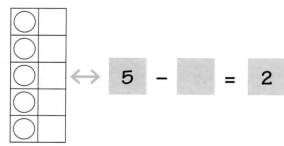

$$5 - \boxed{} = 2$$

Finding subtraction facts from addition facts and vice versa

> I know that 1 plus 5 equals 6. So 6 minus 1 equals 5, and 6 minus 5 equals 1.

A

$6 - 1 =$ ☐ ⟵ $5 + 1 =$ 6 ⟶ $6 - 5 =$ ☐

$5 - 2 =$ ☐ ⟵ $3 + 2 =$ ☐ ⟶ $5 - 3 =$ ☐

$8 - 1 =$ ☐ ⟵ $1 + 7 =$ ☐ ⟶ $8 - 7 =$ ☐

B

$5 - 3 =$ ☐ ⟵ $2 + 3 =$ 5 ⟶ $5 - 2 =$ ☐

$10 - 1 =$ ☐ ⟵ $1 + 9 =$ ☐ ⟶ $10 - 9 =$ ☐

$8 - 1 =$ ☐ ⟵ $7 + 1 =$ ☐ ⟶ $8 - 7 =$ ☐

C

$5 + 5 =$ ☐ ⟵ $10 - 5 =$ 5 ⟶ $10 -$ ☐ $= 5$

$2 + 2 =$ ☐ ⟵ $4 - 2 =$ ☐ ⟶ $4 -$ ☐ $= 2$

$1 + 1 =$ ☐ ⟵ $2 - 1 =$ ☐ ⟶ $2 -$ ☐ $= 1$

Practising basic addition and subtraction facts involving words

8 dogs + 1 dog = ☐ dogs   2 cats + 3 cats = ☐ cats

6 tens − 1 ten = ☐ tens   1 dollar + 7 dollars = 8 ☐

2 cents + 2 cents = ☐ cents   5 buttons + 1 button = 6 ☐

Repeatedly practising 2 + 1 = 3, 3 + 1 = 4 and 1 + 1 = 2

Ⓐ

2	+	1	=	3
	+		=	
	+		=	
	+		=	
	+		=	
1	+	2	=	3
	+		=	
	+		=	
	+		=	
	+		=	
3	–	1	=	
3	–		=	1
3	–	2	=	
3	–		=	2
	–	2	=	1
	–	1	=	2
2	+		=	3
	+	2	=	3
1	+		=	3
	+	1	=	3

Ⓑ

3	+	1	=	4
	+		=	
	+		=	
	+		=	
	+		=	
1	+	3	=	4
	+		=	
	+		=	
	+		=	
	+		=	
4	–	1	=	
4	–		=	1
4	–	3	=	
4	–		=	3
	–	3	=	1
	–	1	=	3
1	+		=	4
	+	3	=	4
1	+		=	4
	+	1	=	4

Ⓒ

1	+	1	=	2
	+		=	
	+		=	
	+		=	
	+		=	
	+		=	
	+		=	
	+		=	
	+		=	
	+		=	
2	–	1	=	
2	–		=	1
2	–	1	=	
2	–		=	1
	–	1	=	1
2	–		=	1
1	+		=	2
	+	1	=	2
1	+		=	2
	+	1	=	2

Repeatedly practising 4 + 1 = 5, 3 + 2 = 5 and 2 + 2 = 4

A

4	+	1	=	5
	+		=	
	+		=	
	+		=	
	+		=	
1	+	4	=	5
	+		=	
	+		=	
	+		=	
	+		=	

5	–	1	=	
5	–		=	1
5	–	4	=	
5	–		=	4
	–	4	=	1
	–	1	=	4

4	+		=	5
	+	1	=	5
1	+		=	5
	+	1	=	5

B

3	+	2	=	5
	+		=	
	+		=	
	+		=	
	+		=	
2	+	3	=	5
	+		=	
	+		=	
	+		=	
	+		=	

5	–	2	=	
5	–		=	2
5	–	3	=	
5	–		=	3
	–	3	=	2
	–	2	=	3

2	+		=	5
	+	3	=	5
2	+		=	5
	+	2	=	5

C

2	+	2	=	4
	+		=	
	+		=	
	+		=	
	+		=	
	+		=	
	+		=	
	+		=	
	+		=	
	+		=	

4	–	2	=	
4	–		=	2
4	–	2	=	
4	–		=	2
	–	2	=	2
4	–		=	2

2	+		=	4
	+	2	=	4
2	+	1	=	4
	+	2	=	4

Repeatedly practising 7 + 1 = 8, 6 + 1 = 7 and 5 + 1 = 6

A	B	C
7 + 1 = 8	6 + 1 = 7	5 + 1 = 6
☐ + ☐ = ☐	☐ + ☐ = ☐	☐ + ☐ = ☐
☐ + ☐ = ☐	☐ + ☐ = ☐	☐ + ☐ = ☐
☐ + ☐ = ☐	☐ + ☐ = ☐	☐ + ☐ = ☐
☐ + ☐ = ☐	☐ + ☐ = ☐	☐ + ☐ = ☐
1 + 7 = 8	1 + 6 = 7	1 + 5 = 6
☐ + ☐ = ☐	☐ + ☐ = ☐	☐ + ☐ = ☐
☐ + ☐ = ☐	☐ + ☐ = ☐	☐ + ☐ = ☐
☐ + ☐ = ☐	☐ + ☐ = ☐	☐ + ☐ = ☐
☐ + ☐ = ☐	☐ + ☐ = ☐	☐ + ☐ = ☐
8 − 1 = ☐	7 − 1 = ☐	6 − 1 = ☐
8 − ☐ = 1	7 − ☐ = 1	6 − ☐ = 1
8 − 7 = ☐	7 − 6 = ☐	6 − 5 = ☐
8 − ☐ = 7	7 − ☐ = 6	6 − ☐ = 5
☐ − 7 = 1	☐ − 6 = 1	☐ − 5 = 1
☐ − 1 = 7	☐ − 1 = 6	☐ − 1 = 5
7 + ☐ = 8	6 + ☐ = 7	5 + ☐ = 6
☐ + 7 = 8	☐ + 6 = 7	☐ + 5 = 6
1 + ☐ = 8	1 + ☐ = 7	1 + ☐ = 6
☐ + 1 = 8	☐ + 1 = 7	☐ + 1 = 6

Repeatedly practising 8 + 1 = 9, 6 + 1 = 7 and 9 + 1 = 10

A

$8 + 1 = 9$

$\square + \square = \square$

$\square + \square = \square$

$\square + \square = \square$

$\square + \square = \square$

$1 + 8 = 9$

$\square + \square = \square$

$\square + \square = \square$

$\square + \square = \square$

$\square + \square = \square$

$9 - 1 = \square$

$9 - \square = 1$

$9 - 8 = \square$

$9 - \square = 8$

$\square - 8 = 1$

$\square - 1 = 8$

$8 + \square = 9$

$\square + 8 = 9$

$1 + \square = 9$

$\square + 1 = 9$

B

$6 + 1 = 7$

$\square + \square = \square$

$\square + \square = \square$

$\square + \square = \square$

$\square + \square = \square$

$1 + 6 = 7$

$\square + \square = \square$

$\square + \square = \square$

$\square + \square = \square$

$\square + \square = \square$

$7 - 1 = \square$

$7 - \square = 1$

$7 - 6 = \square$

$7 - \square = 6$

$\square - 6 = 1$

$\square - 1 = 6$

$6 + \square = 7$

$\square + 6 = 7$

$1 + \square = 7$

$\square + 1 = 7$

C

$9 + 1 = 10$

$\square + \square = \square$

$\square + \square = \square$

$\square + \square = \square$

$\square + \square = \square$

$1 + 9 = 10$

$\square + \square = \square$

$\square + \square = \square$

$\square + \square = \square$

$\square + \square = \square$

$10 - 1 = \square$

$10 - \square = 1$

$10 - 9 = \square$

$10 - \square = 9$

$\square - 9 = 1$

$\square - 1 = 9$

$9 + \square = 10$

$\square + 9 = 10$

$1 + \square = 10$

$\square + 1 = 10$

Unit 1

13

Extras: practising and applying basic addition and subtraction facts

A

☐ − 3 = 2 ⟵ 2 + 3 = ☐ ⟶ 5 − ☐ = 2
☐ − 2 = 1 ⟵ 1 + 2 = ☐ ⟶ 3 − ☐ = 1
☐ − 7 = 1 ⟵ 1 + 7 = ☐ ⟶ 8 − ☐ = 7
☐ − 2 = 3 ⟵ 3 + 2 = ☐ ⟶ 5 − ☐ = 3
☐ − 6 = 1 ⟵ 6 + 1 = ☐ ⟶ 7 − ☐ = 1
☐ − 8 = 1 ⟵ 8 + 1 = ☐ ⟶ 9 − ☐ = 1

B 5 books + ☐ book = 6 books   1 plum + ☐ plums = 8 plums

2 cents + ☐ cents = 5 cents   1 apple + ☐ apples = 6 apples

1 girl + ☐ girls = 7 girls   ☐ pears + 3 pears = 5 pears

1 tree + ☐ trees = 9 trees   ☐ fingers + 1 finger = 5 fingers

C 8 thousand dollars + 1 ☐ dollars = 9 thousand dollars

3 hundred dollars + 2 ☐ dollars = 5 hundred dollars

4 million seeds + 1 ☐ seeds = 5 million seeds

1 packet of rice + 7 ☐ = 8 packets of rice

D Sarah has 3 cookies in a jar under her bed. While she is asleep her mother puts some more cookies in the jar. In the morning Sarah wakes up and counts 5 cookies in the jar. How many cookies did her mother put in the jar at night?

Write the answer: ☐ cookies

Counting forwards and backwards between 1 and 30

1	2	3		5		7			10

6	7	8		10		12		14	

10	9	8		6					1

14	13	12			9	8			5

20		18	17		15		13		11

30	29		27	26					21

11	12			15		17			20

Adding with five on ten frames

I add with five on a ten frame.

A 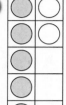 → 5 + 2 = ☐

→ 5 + 3 = ☐

→ 4 + 5 = ☐

→ ☐ + 5 = 8

→ 5 + ☐ = 10

→ ☐ + ☐ = 7

↔ 5 + ☐ = 9

↔ 4 + ☐ = 9

To take away five, I colour in five circles with a dark colour. I see how many circles are left.

$$9 - 5 = 4$$

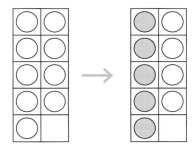

A

 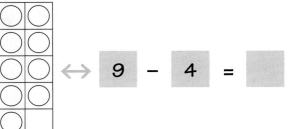

$$9 - 4 = \boxed{}$$

$$9 - \boxed{} = 5$$

$$6 - \boxed{} = 1$$

$$5 - \boxed{} = 3$$

B

$$8 - 5 = \boxed{}$$

$$7 - \boxed{} = 2$$

$$8 - \boxed{} = 3$$

$$8 - \boxed{} = 5$$

Unit 2

Basic addition facts lead to subtraction facts

I know that 4 plus 5 equals 9. So 9 minus 4 equals 5, and 9 minus 5 equals 4.

Ⓐ   9 − 4 = ☐   ⟵   5 + 4 = ☐   ⟶   9 − 5 = ☐

5 + 2 = ☐   ⟵   7 − 2 = ☐   ⟶   2 + 5 = ☐

8 − 3 = ☐   ⟵   3 + 5 = ☐   ⟶   8 − 5 = ☐

Ⓑ   5 − 3 = ☐   ⟵   2 + 3 = ☐   ⟶   5 − 2 = ☐

1 + 5 = ☐   ⟵   6 − 1 = ☐   ⟶   5 + 1 = ☐

7 − 2 = ☐   ⟵   5 + 2 = ☐   ⟶   7 − 5 = ☐

Ⓒ   10 − ☐ = 5   ⟵   5 + 5 = ☐   ⟶   10 − 5 = ☐

4 + ☐ = 9   ⟵   9 − 4 = ☐   ⟶   5 + 4 = ☐

7 − ☐ = 2   ⟵   2 + 5 = ☐   ⟶   7 − 2 = ☐

Ⓓ Michaela has $8. She spends $5 at the shop. How much money does she have left?

Write the answer: $ ☐

Ⓔ Laurence has nine apples. He gives away four apples to his friends. How many apples does Laurence have left?

Write the answer: ☐ apples

Ⓕ Samantha has 5 bananas and she eats some of them. Now she has 1 banana left. How many bananas did she eat?

Write the answer: ☐ bananas

Repeatedly practising 1 + 4 = 5, 5 + 1 = 6 and 1 + 6 = 7

A

$1 + 4 = 5$

$\square + \square = \square$

$\square + \square = \square$

$\square + \square = \square$

$\square + \square = \square$

$4 + 1 = 5$

$\square + \square = \square$

$\square + \square = \square$

$\square + \square = \square$

$\square + \square = \square$

$5 - 1 = \square$

$5 - \square = 1$

$1 + \square = 5$

$\square + 4 = 5$

$5 - 4 = \square$

$\square - 1 = 4$

$1 + 4 = \square$

B

$5 + 1 = 6$

$\square + \square = \square$

$\square + \square = \square$

$\square + \square = \square$

$\square + \square = \square$

$1 + 5 = 6$

$\square + \square = \square$

$\square + \square = \square$

$\square + \square = \square$

$\square + \square = \square$

$6 - 1 = \square$

$6 - \square = 1$

$5 + \square = 6$

$\square + 1 = 6$

$6 - 1 = \square$

$\square - 5 = 1$

$5 + 1 = \square$

C

$1 + 6 = 7$

$\square + \square = \square$

$\square + \square = \square$

$\square + \square = \square$

$\square + \square = \square$

$6 + 1 = 7$

$\square + \square = \square$

$\square + \square = \square$

$\square + \square = \square$

$\square + \square = \square$

$7 - 1 = \square$

$7 - \square = 1$

$1 + \square = 7$

$\square + 1 = 7$

$7 - 1 = \square$

$\square - 6 = 1$

$1 + 6 = \square$

Unit 2

A

5 + 2 = 7

☐ + ☐ = ☐

☐ + ☐ = ☐

☐ + ☐ = ☐

☐ + ☐ = ☐

2 + 5 = 7

☐ + ☐ = ☐

☐ + ☐ = ☐

☐ + ☐ = ☐

☐ + ☐ = ☐

7 − 2 = ☐

7 − ☐ = 2

2 + ☐ = 7

☐ + 5 = 7

7 − 2 = ☐

☐ − 5 = 2

5 + 2 = ☐

B

5 + 3 = 8

☐ + ☐ = ☐

☐ + ☐ = ☐

☐ + ☐ = ☐

☐ + ☐ = ☐

3 + 5 = 8

☐ + ☐ = ☐

☐ + ☐ = ☐

☐ + ☐ = ☐

☐ + ☐ = ☐

8 − 3 = ☐

8 − ☐ = 3

5 + ☐ = 8

☐ + 3 = 8

8 − 3 = ☐

☐ − 3 = 5

3 + 5 = ☐

C

5 + 4 = 9

☐ + ☐ = ☐

☐ + ☐ = ☐

☐ + ☐ = ☐

☐ + ☐ = ☐

4 + 5 = 9

☐ + ☐ = ☐

☐ + ☐ = ☐

☐ + ☐ = ☐

☐ + ☐ = ☐

9 − 4 = ☐

9 − ☐ = 4

5 + ☐ = 9

☐ + 4 = 9

9 − 4 = ☐

☐ − 4 = 5

4 + 5 = ☐

A

$2 + 3 = 5$

$\square + \square = \square$

$\square + \square = \square$

$\square + \square = \square$

$\square + \square = \square$

$3 + 2 = 5$

$\square + \square = \square$

$\square + \square = \square$

$\square + \square = \square$

$\square + \square = \square$

$5 - 2 = \square$

$5 - \square = 2$

$2 + \square = 5$

$\square + 2 = 5$

$5 - 2 = \square$

$\square - 3 = 2$

$2 + 3 = \square$

B

$2 + 5 = 7$

$\square + \square = \square$

$\square + \square = \square$

$\square + \square = \square$

$\square + \square = \square$

$5 + 2 = 7$

$\square + \square = \square$

$\square + \square = \square$

$\square + \square = \square$

$\square + \square = \square$

$7 - 2 = \square$

$7 - \square = 2$

$2 + \square = 7$

$\square + 2 = 7$

$7 - 2 = \square$

$\square - 2 = 5$

$2 + 5 = \square$

C

$5 + 5 = 10$

$\square + \square = \square$

$\square + \square = \square$

$\square + \square = \square$

$\square + \square = \square$

$\square + \square = \square$

$\square + \square = \square$

$\square + \square = \square$

$\square + \square = \square$

$10 - 5 = \square$

$10 - \square = 5$

$5 + \square = 10$

$\square + 5 = 10$

$10 - 5 = \square$

$\square - 5 = 5$

$5 + 5 = \square$

Unit 2

21

In numbers, 'teen' means ten

Thir is another way to say three.

Teen is another way to say ten.

Fif is another way to say five.

A fourteen = ten plus [four]

nineteen = ten plus [ ]

thirteen = ten plus [ ]

B fifteen = ten plus [ ]

eighteen = ten plus [ ]

sixteen = ten plus [ ]

C The word for ten plus two is [ ]

The word for ten plus one is [ ]

D ten plus seven = [seventeen]

ten plus five = [ ]

ten plus four = [ ]

three plus ten = [ ]

ten plus two = [ ]

five plus ten = [ ]

E ten plus six = [ ]

ten plus eight = [ ]

ten plus nine = [ ]

one plus ten = [ ]

two plus ten = [ ]

nine plus ten = [ ]

F Sala has six jellybeans. Her grandmother gives her another ten jellybeans. How many jellybeans does Sala have now?

Write the answer: [ ] jellybeans

G Jerry buys two toy cars. The red car costs ten dollars and the green car costs nine dollars. How much money does Jerry pay the shopkeeper?

Write the answer: [ ] dollars

More practice with teen words

A

7 + ten = seventeen

5 + ten = _____

4 + ten = _____

ten + 3 = _____

B

six + 10 = sixteen

eight + 10 = _____

10 + nine = _____

2 + ten = _____

Linking place value to the teen words

11 eleven	14 fourteen	17 seventeen
12 twelve	15 fifteen	18 eighteen
13 thirteen	16 sixteen	19 nineteen

Tens	Ones				
1	8	→	1 ten + 8 ones	→	eighteen

Tens	Ones				
1	6	→	___ ten + ___ ones	→	_____

Tens	Ones				
1	4	→	___ ten + ___ ones	→	_____

Tens	Ones				
1	5	→	1 ten + ___ ones	→	_____

Tens	Ones				
1	3	→	1 ten + ___ ones	→	_____

Unit 2

Linking teen words to place value

A

						Tens	Ones
twelve	→	1	ten +	2	ones →	1	2

						Tens	Ones
thirteen	→		ten +		ones →		

						Tens	Ones
nineteen	→		ten +		ones →		

						Tens	Ones
	←		ten +		ones ←	1	6

B

Tens	Ones		
1	7	→	17

Tens	Ones		
		←	19

C

Tens	Ones		
1	4	→	

Tens	Ones		
		←	15

D

sixteen	→	16
nineteen	→	
	←	13

E

fourteen	→	
twelve	→	
	←	18

Linking ten frames to teen words

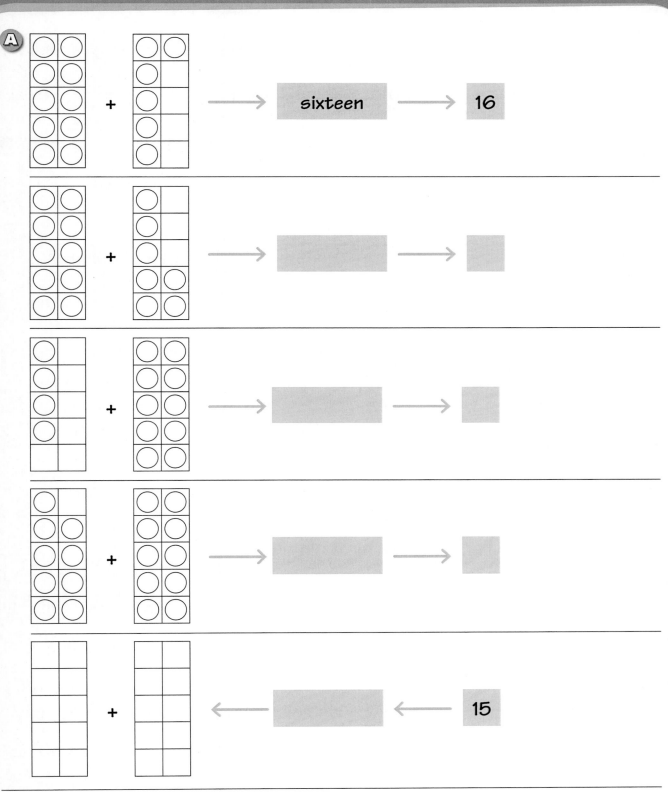

A

+ ⟶ sixteen ⟶ 16

+ ⟶ [ ] ⟶ [ ]

+ ⟶ [ ] ⟶ [ ]

+ ⟶ [ ] ⟶ [ ]

+ ⟵ [ ] ⟵ 15

Unit 2

B Mela buys two books. One book costs three dollars and the other book
costs ten dollars. How much money does Mela pay for the books?

Write the answer: [ ] dollars

A 5 cents + 2 cents = ☐ cents

4 tens + 5 tens = ☐ tens

6 tens – 1 ten = ☐ tens

9 cans – 1 can = ☐ cans

B 5 buttons + 3 buttons = ☐ buttons

10 dollars + 2 dollars = ☐ dollars

7 dollars – 2 dollars = ☐ dollars

5 cents – 1 cent = 4 ☐

C 1 thousand dollars + 7 thousand dollars = ☐ thousand dollars

8 thousand dollars – 5 thousand dollars = ☐ ☐ dollars

5 hundred dollars + 3 hundred dollars = ☐ hundred dollars

7 million + 10 million = ☐ ☐

5 tomatoes – 4 tomatoes = ☐ ☐

17 dollars – 7 dollars + 2 dollars = ☐ dollars

5 fish + 5 fish + 1 fish = ☐ ☐

D Melissa has $13. She spends $3. How much money does she have left?

Write the answer: $ ☐

E Marie buys two teddy bears. The big teddy bear costs ten dollars and the little teddy bear costs seven dollars. How much money does Marie pay the shopkeeper?

Write the answer: ☐ dollars

Extras: more revision of basic addition and subtraction facts

A

9 − [ ] = 5 ← 5 + 4 = [ ] → [ ] − 4 = 5

8 − [ ] = 5 ← 3 + 5 = [ ] → [ ] − 5 = 3

7 − [ ] = 2 ← 2 + 5 = [ ] → [ ] − 5 = 2

10 − [ ] = 5 ← 5 + 5 = [ ] → [ ] − 5 = 5

B

14 − [ ] = 10 ← 10 + 4 = [ ] → [ ] − 4 = 10

11 − [ ] = 1 ← 1 + 10 = [ ] → [ ] − 10 = 1

12 − [ ] = 10 ← 2 + 10 = [ ] → [ ] − 2 = 10

19 − [ ] = 9 ← 10 + 9 = [ ] → [ ] − 10 = 9

C

5 books + [ ] books = 9 books   5 plums + [ ] plums = 8 plums

12 trees − [ ] trees = 10 trees   [ ] cents − 4 cents = 10 cents

5 books + [ ] books = 9 books   5 pens + [ ] pens = 15 pens

19 trees − [ ] trees = 10 trees   [ ] cents − 10 cents = 2 cents

D

5 monkeys + [ ] monkeys = 10 monkeys

5 billion people + [ ] [ ] people = 7 billion people

E Miles has sixteen cookies in a jar under his bed. When he is asleep his little brother comes in to his room and eats some cookies. In the morning Miles counts his cookies. He finds he has only six left. How many cookies did his little brother eat?

Write the answer: [ ] cookies

Unit 2

| 31 | 32 | 33 | | 35 | | | | | 40 |

| 56 | 57 | 58 | | 60 | | | | | 65 |

| 40 | 39 | 38 | | 36 | | | | | 31 |

| 74 | 73 | | 71 | | 69 | | | | 65 |

| 99 | | 97 | | 95 | | 93 | | 91 | |

| 53 | 52 | | 50 | | 48 | | 46 | | |

| 57 | 58 | | 60 | | 62 | | | | 66 |

Using ten frames to add numbers

I look at the pictures and know how to add.
I do need not to count.

A 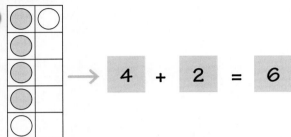 → 4 + 2 = 6

B 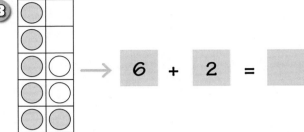 → 6 + 2 = ☐

→ 2 + 7 = ☐

→ 2 + 8 = ☐

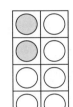 → 3 + ☐ = 6

 → 3 + ☐ = 9

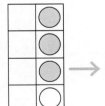 → 3 + ☐ = 10

→ 6 + ☐ = 9

Unit 3

A

4 + 4 = 8 →

4 + 6 = ↔

+ 4 = 9 ↔

B

7 + 2 = 9 →

7 + = 10 ↔

2 + 6 = ↔

C Geraldine needs 10 dollars to buy a toy. She has 6 dollars. How much more money does she have to save to buy the toy?

Write the answer: dollars

D Six children in Mr Tose's class order orange drinks for lunch. Four children in Ms Watt's class order orange drinks. How many orange drinks are ordered altogether?

Write the answer:  drinks

Subtraction with numbers from 1 to 9 on ten frames

To take away 5, I colour in five circles with a dark colour.

$$9 - 5 = 4$$

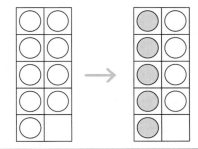

A $9 - 6 = \boxed{}$

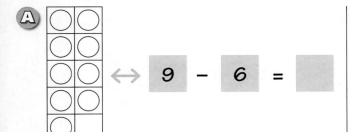

$9 - \boxed{} = 6$

$6 - \boxed{} = 2$

$5 - \boxed{} = 3$

B $8 - 4 = \boxed{}$

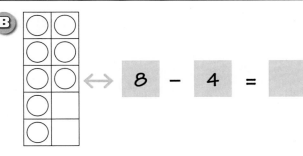

$7 - \boxed{} = 3$

$8 - \boxed{} = 3$

$8 - \boxed{} = 2$

Unit 3

Basic addition facts lead to subtraction facts

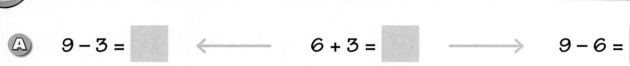

I know 3 plus 6 equals 9. So 9 minus 6 equals 3 and 9 minus 3 equals 6.

A

9 − 3 = ☐ ⟵ 6 + 3 = ☐ ⟶ 9 − 6 = ☐

7 + 2 = ☐ ⟵ 9 − 2 = ☐ ⟶ 2 + 7 = ☐

10 − 2 = ☐ ⟵ 2 + 8 = ☐ ⟶ 10 − 8 = ☐

B

8 − 6 = ☐ ⟵ 2 + 6 = ☐ ⟶ 8 − 2 = ☐

7 + 3 = ☐ ⟵ 10 − 3 = ☐ ⟶ 3 + 7 = ☐

7 − 4 = ☐ ⟵ 3 + 4 = ☐ ⟶ 7 − 3 = ☐

C

10 − ☐ = 4 ⟵ 4 + 6 = ☐ ⟶ 10 − ☐ = 6

4 + ☐ = 9 ⟵ 5 + 4 = ☐ ⟶ 5 + ☐ = 9

6 − ☐ = 4 ⟵ 4 + 2 = ☐ ⟶ 6 − ☐ = 2

D In Melissa's fruit stall, there are ten apples when the stall opens. She sells seven apples. How many apples does Melissa have left?

Write the answer: ☐ apples

E John has to cook ten potatoes for dinner but he only has seven potatoes. How many more potatoes does John need?

Write the answer: ☐ potatoes

F Meredith saves coins. Meredith adds four more coins to her collection on Tuesday and now she has ten coins. How many coins did Meredith have on Monday?

Write the answer: ☐ coins

Repeatedly practising 3 + 3 = 6, 3 + 4 = 7 and 2 + 4 = 6

A

3	+ 3	=	6
	+	=	
	+	=	
	+	=	
	+	=	
	+	=	
	+	=	
	+	=	
	+	=	
	+	=	

6 − 3 =

6 − = 3

3 + = 6

+ 3 = 6

6 − 3 =

− 3 = 3

3 + 3 =

B

3	+ 4	=	7
	+	=	
	+	=	
	+	=	
	+	=	
4	+ 3	=	7
	+	=	
	+	=	
	+	=	
	+	=	

7 − 3 =

7 − = 3

4 + = 7

+ 3 = 7

7 − 3 =

− 4 = 3

4 + 3 =

C

2	+ 4	=	6
	+	=	
	+	=	
	+	=	
	+	=	
4	+ 2	=	6
	+	=	
	+	=	
	+	=	
	+	=	

6 − 2 =

6 − = 2

4 + = 6

+ 2 = 6

6 − 2 =

− 4 = 2

4 + 2 =

Unit 3

A

4 + 3 = 7

☐ + ☐ = ☐

☐ + ☐ = ☐

☐ + ☐ = ☐

☐ + ☐ = ☐

3 + 4 = 7

☐ + ☐ = ☐

☐ + ☐ = ☐

☐ + ☐ = ☐

☐ + ☐ = ☐

7 − 3 = ☐

7 − ☐ = 3

3 + ☐ = 7

☐ + 4 = 7

7 − 3 = ☐

☐ − 3 = 4

3 + 4 = ☐

B

5 + 2 = 7

☐ + ☐ = ☐

☐ + ☐ = ☐

☐ + ☐ = ☐

☐ + ☐ = ☐

2 + 5 = 7

☐ + ☐ = ☐

☐ + ☐ = ☐

☐ + ☐ = ☐

☐ + ☐ = ☐

7 − 2 = ☐

7 − ☐ = 2

5 + ☐ = 7

☐ + 2 = 7

7 − 2 = ☐

☐ − 5 = 2

5 + 2 = ☐

C

5 + 5 = 10

☐ + ☐ = ☐

☐ + ☐ = ☐

☐ + ☐ = ☐

☐ + ☐ = ☐

☐ + ☐ = ☐

☐ + ☐ = ☐

☐ + ☐ = ☐

☐ + ☐ = ☐

10 − 5 = ☐

10 − ☐ = 5

5 + ☐ = 10

☐ + 5 = 10

10 − 5 = ☐

☐ − 5 = 5

5 + 5 = ☐

Repeatedly practising 2 + 6 = 8, 3 + 6 = 9 and 2 + 7 = 9

A

2 + 6 = 8

☐ + ☐ = ☐

☐ + ☐ = ☐

☐ + ☐ = ☐

☐ + ☐ = ☐

6 + 2 = 8

☐ + ☐ = ☐

☐ + ☐ = ☐

☐ + ☐ = ☐

☐ + ☐ = ☐

8 − 2 = ☐

8 − ☐ = 2

6 + ☐ = 8

☐ + 6 = 8

8 − 6 = ☐

☐ − 2 = 6

2 + 6 = ☐

B

3 + 6 = 9

☐ + ☐ = ☐

☐ + ☐ = ☐

☐ + ☐ = ☐

☐ + ☐ = ☐

6 + 3 = 9

☐ + ☐ = ☐

☐ + ☐ = ☐

☐ + ☐ = ☐

☐ + ☐ = ☐

9 − 3 = ☐

9 − ☐ = 3

6 + ☐ = 9

☐ + 3 = 9

9 − 3 = ☐

☐ − 6 = 3

6 + 3 = ☐

C

2 + 7 = 9

☐ + ☐ = ☐

☐ + ☐ = ☐

☐ + ☐ = ☐

☐ + ☐ = ☐

7 + 2 = 9

☐ + ☐ = ☐

☐ + ☐ = ☐

☐ + ☐ = ☐

☐ + ☐ = ☐

9 − 2 = ☐

9 − ☐ = 2

7 + ☐ = 9

☐ + 2 = 9

9 − 2 = ☐

☐ − 7 = 2

7 + 2 = ☐

Repeatedly practising 2 + 8 = 10, 3 + 7 = 10 and 4 + 6 = 10

A

2 + 8 = 10

☐ + ☐ = ☐

☐ + ☐ = ☐

☐ + ☐ = ☐

☐ + ☐ = ☐

8 + 2 = 10

☐ + ☐ = ☐

☐ + ☐ = ☐

☐ + ☐ = ☐

☐ + ☐ = ☐

10 – 2 = ☐

10 – ☐ = 2

8 + ☐ = 10

☐ + 8 = 10

10 – 2 = ☐

☐ – 8 = 2

2 + 8 = ☐

B

3 + 7 = 10

☐ + ☐ = ☐

☐ + ☐ = ☐

☐ + ☐ = ☐

☐ + ☐ = ☐

7 + 3 = 10

☐ + ☐ = ☐

☐ + ☐ = ☐

☐ + ☐ = ☐

☐ + ☐ = ☐

10 – 3 = ☐

10 – ☐ = 3

7 + ☐ = 10

☐ + 3 = 10

10 – 3 = ☐

☐ – 7 = 3

7 + 3 = ☐

C

4 + 6 = 10

☐ + ☐ = ☐

☐ + ☐ = ☐

☐ + ☐ = ☐

☐ + ☐ = ☐

6 + 4 = 10

☐ + ☐ = ☐

☐ + ☐ = ☐

☐ + ☐ = ☐

☐ + ☐ = ☐

10 – 4 = ☐

10 – ☐ = 4

6 + ☐ = 10

☐ + 4 = 10

10 – 6 = ☐

☐ – 4 = 6

4 + 6 = ☐

Number words ending in '-ty' is a code meaning 'tens'

> Forty is another way to say four tens.

> Sixty is another way to say six tens.

> Two tens is another way to say twenty.

A
forty = four tens

eighty = ☐ tens

fifty = ☐ tens

thirty = ☐ tens

B
sixty = ☐ tens

ninety = ☐ tens

seventy = ☐ tens

twenty = ☐ tens

C
seven tens = seventy

eight tens = ☐

three tens = ☐

four tens = ☐

D
six tens = ☐

two tens = ☐

five tens = ☐

nine tens = ☐

Unit 3

Turning two-digit number words into their place-value form

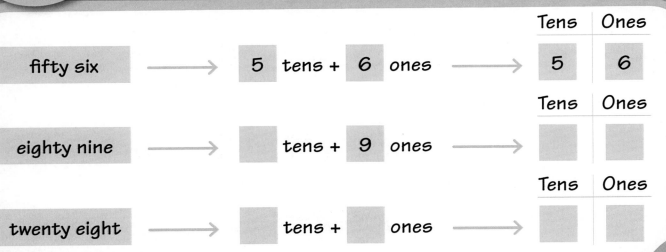

			Tens	Ones
fifty six	→	5 tens + 6 ones →	5	6

			Tens	Ones
eighty nine	→	☐ tens + 9 ones →	☐	☐

			Tens	Ones
twenty eight	→	☐ tens + ☐ ones →	☐	☐

Turning two-digit numbers into their word form

Tens	Ones				
7	4	→	7 tens + 4 ones	→	seventy four

Tens	Ones				
3	7	→	___ tens + ___ ones	→	___

Tens	Ones				
2	0	→	___ tens + ___ ones	→	___

Tens	Ones				
6	6	→	___ tens + ___ ones	→	___

Turning objects grouped in tens into their word form

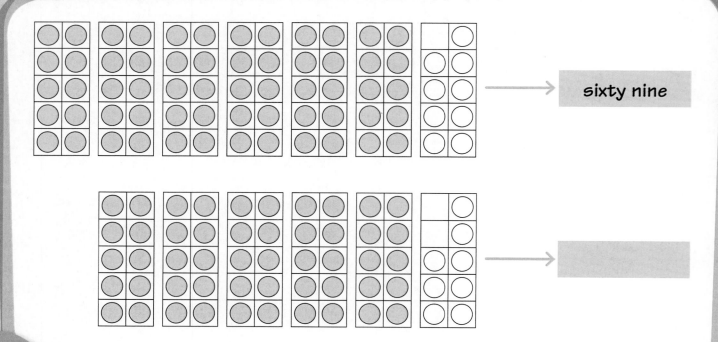

→ sixty nine

Converting numbers into a place-value table, two-digit form and words

A

Tens	Ones
8	7

⟶ 87

Tens	Ones
6	2

⟶

Tens	Ones

⟵ 29

Tens	Ones

⟵ 90

B

Tens	Ones
3	4

⟶

Tens	Ones
7	0

⟶

Tens	Ones

⟵ 55

Tens	Ones

⟵ 62

C

seventy three ⟶ 73

twenty seven ⟶

fifty ⟶

⟵ 46

⟵ 13

⟵ 71

D

forty eight ⟶

twelve ⟶

nineteen ⟶

⟵ 18

⟵ 80

⟵ 42

Unit 3

Unit 3

A 7 cents + 2 cents = ☐ cents

4 tens + 5 tens = ☐ tens

6 tens − 4 tens = ☐ tens

9 cans − 7 cans = ☐ cans

B 8 buttons + 2 buttons = ☐ buttons

10 dollars + 2 dollars = 12 ☐

8 dollars − 2 dollars = 6 ☐

5 cents − 3 cents = ☐ ☐

C 9 thousand dollars − 5 thousand dollars = ☐ ☐ dollars

7 hundred dollars + 2 hundred dollars = ☐ ☐ dollars

2 thousand dollars + ☐ thousand dollars = 8 thousand dollars

6 metres − 1 metre = ☐ metres

D The school is raising funds by organising a walkathon. Mele gives her teacher ten dollars that she raised on the walk. Ah See gives the teacher eight dollars. How much money do they give the teacher altogether?

Write the answer: ☐ dollars

E There are eighteen apples at the school tuck shop. At lunchtime 8 children buy an apple each. How many apples are left at the tuck shop at the end of lunchtime?

Write the answer: ☐ apples

F Maureen buys ten oranges at the supermarket and eats some of them on the way home. When she gets home she has only seven oranges. How many oranges did Maureen eat?

Write the answer: ☐ oranges

Extras: revising basic addition and subtraction facts

A 14 − ☐ = 4 ⟵ 10 + 4 = ☐ ⟶ ☐ − 4 = 10

15 − ☐ = 5 ⟵ 10 + 5 = ☐ ⟶ ☐ − 5 = 10

12 − ☐ = 2 ⟵ 2 + 10 = ☐ ⟶ ☐ − 2 = 10

19 − ☐ = 9 ⟵ 10 + 9 = ☐ ⟶ ☐ − 9 = 10

B 10 − ☐ = 4 ⟵ 6 + 4 = ☐ ⟶ ☐ − 4 = 6

10 − ☐ = 2 ⟵ 2 + 8 = ☐ ⟶ ☐ − 8 = 2

10 − ☐ = 6 ⟵ 4 + 6 = ☐ ⟶ ☐ − 6 = 4

10 − ☐ = 7 ⟵ 7 + 3 = ☐ ⟶ ☐ − 7 = 3

C 1 book + ☐ books = 6 books

12 trees − ☐ trees = 10 trees

$19 − $☐ = $10

$13 − $☐ = $3

$10 − $☐ = $2

D 5 plums + ☐ plums = 8 plums

☐ cents − 4 cents = 10 ☐

$☐ − $10 = $3

$☐ − $5 = $4

$☐ − $6 = $10

E At the end of work one day a farmer notices that 6 eggs have hatched into chicks. The next morning, the farmer counts the chicks and finds there are now ten. How many eggs hatched during the night?

Write the answer: ☐ eggs

| 450 | 451 | 452 | | 454 | | | | | 459 |

| 817 | 818 | 819 | | | 822 | | | | 826 |

| 904 | 903 | 902 | | | 899 | | | | 895 |

| 303 | 302 | 301 | | | | | | | 294 |

| 675 | 676 | | 678 | | | | | | |

| 685 | | 687 | | 689 | | | | | 694 |

| 617 | | 619 | | 621 | | | | | 626 |

| 627 | 628 | | 630 | | | | | | |

| 637 | | 639 | | | | | | | 646 |

Deriving the basic addition facts by adding one or two

I need to know all the pairs of numbers that add up to ten.

A 5 + 5 = 10 ⟶ 6 + 5 = ☐

4 + 6 = 10 ⟶ 4 + 7 = ☐

8 + 2 = 10 ⟶ 8 + 3 = ☐

2 + 8 = 10 ⟶ 3 + 8 = ☐

B 7 + 3 = 10 ⟶ 7 + 5 = ☐

6 + 4 = 10 ⟶ 6 + 6 = ☐

1 + 9 = 10 ⟶ 3 + 9 = ☐

7 + 3 = 10 ⟶ 9 + 3 = ☐

Using the basic derived addition facts with words and stories

A 3 dogs + 9 dogs = ☐ dogs

7 cents + 5 cents = ☐ cents

6 plums + 5 plums = ☐ plums

B 7 cats + 4 cats = ☐ cats

6 lions + 9 lions = 15 ☐

4 bats + 8 bats = ☐ ☐

C Write a story for 8 dollars + 4 dollars.

_____

_____

Write your question: _____

Answer: ☐ dollars

Equation: ☐ + ☐ = ☐

Unit 4

Deriving more basic addition facts by adding one or two

7 + 4 = ☐   ⟵   7 + 3 = ☐   ⟶   8 + 3 = ☐

4 + 8 = ☐   ⟵   2 + 8 = ☐   ⟶   2 + 10 = ☐

5 + 6 = ☐   ⟵   5 + 5 = ☐   ⟶   6 + 5 = ☐

3 + 9 = ☐   ⟵   3 + 7 = ☐   ⟶   5 + 7 = ☐

Deriving the basic addition facts by adding three or more

Ⓐ 5 + 5 = 10 ⟶ 8 + 5 = ☐

4 + 6 = 10 ⟶ 4 + 9 = ☐

8 + 2 = 10 ⟶ 8 + 5 = ☐

2 + 8 = 10 ⟶ 6 + 8 = ☐

Ⓑ 7 + 3 = 10 ⟶ 7 + 6 = ☐

6 + 4 = 10 ⟶ 9 + 4 = ☐

1 + 9 = 10 ⟶ 6 + 9 = ☐

7 + 3 = 10 ⟶ 7 + 7 = ☐

Using the basic derived addition facts with words

Ⓐ 6 figs + 6 figs = ☐ figs

7 cents + 7 cents = ☐ cents

7 plums + 5 plums = ☐ plums

Ⓑ 7 pigs + 4 pigs = ☐ pigs

6 loaves + 7 loaves = ☐ loaves

5 doors + 8 doors = ☐ doors

More deriving the basic addition facts by adding three or more

7 + 6 = ☐	←	7 + 3 = ☐	→	10 + 3 = ☐
5 + 8 = ☐	←	2 + 8 = ☐	→	2 + 11 = ☐
5 + 9 = ☐	←	5 + 5 = ☐	→	9 + 5 = ☐
7 + 7 = ☐	←	3 + 7 = ☐	→	3 + 11 = ☐

Addition facts lead to subtraction facts

13 − 9 = ☐	←	4 + 9 = ☐	→	13 − 4 = ☐
11 − 9 = ☐	←	2 + 9 = ☐	→	11 − 2 = ☐
13 − 7 = ☐	←	6 + 7 = ☐	→	13 − 6 = ☐
13 − 9 = ☐	←	4 + 9 = ☐	→	13 − 4 = ☐
16 − 9 = ☐	←	7 + 9 = ☐	→	16 − 7 = ☐

Unit 4

Applying the basic facts to adding and subtracting three numbers

A
4 + 4 + 8 = ☐
5 + 7 + 3 = ☐
3 + 4 + 8 = ☐

B
3 + 8 − 4 = ☐
8 + 8 − 4 = ☐
7 + 8 − 4 = ☐

C
8 + 7 − 6 = ☐
5 + 8 − 7 = ☐
7 + 8 − 9 = ☐

A

5 + 6 = 11

☐ + ☐ = ☐

☐ + ☐ = ☐

☐ + ☐ = ☐

☐ + ☐ = ☐

6 + 5 = 11

☐ + ☐ = ☐

☐ + ☐ = ☐

☐ + ☐ = ☐

☐ + ☐ = ☐

11 – 5 = ☐

11 – ☐ = 6

6 + ☐ = 11

☐ + 5 = 11

11 – 6 = ☐

☐ – 5 = 6

5 + 6 = ☐

B

6 + 6 = 12

☐ + ☐ = ☐

☐ + ☐ = ☐

☐ + ☐ = ☐

☐ + ☐ = ☐

☐ + ☐ = ☐

☐ + ☐ = ☐

☐ + ☐ = ☐

☐ + ☐ = ☐

☐ + ☐ = ☐

12 – 6 = ☐

12 – ☐ = 6

6 + ☐ = 12

☐ + 6 = 12

12 – 6 = ☐

☐ – 6 = 6

6 + 6 = ☐

C

7 + 6 = 13

☐ + ☐ = ☐

☐ + ☐ = ☐

☐ + ☐ = ☐

☐ + ☐ = ☐

6 + 7 = 13

☐ + ☐ = ☐

☐ + ☐ = ☐

☐ + ☐ = ☐

☐ + ☐ = ☐

13 – 6 = ☐

13 – ☐ = 6

7 + ☐ = 13

☐ + 6 = 13

13 – 6 = ☐

☐ – 7 = 6

7 + 6 = ☐

Repeatedly practising 5 + 7 = 12, 6 + 7 = 13 and 7 + 8 = 15

A

5	+	7	=	12
	+		=	
	+		=	
	+		=	
	+		=	

7	+	5	=	12
	+		=	
	+		=	
	+		=	
	+		=	

12	−	7	=	
12	−		=	5

7	+		=	12
	+	5	=	12

12	−	7	=	
	−	7	=	5

5	+	7	=	

B

6	+	7	=	13
	+		=	
	+		=	
	+		=	
	+		=	

7	+	6	=	13
	+		=	
	+		=	
	+		=	
	+		=	

13	−	6	=	
13	−		=	6

7	+		=	13
	+	6	=	13

13	−	6	=	
	−	7	=	6

7	+	6	=	

C

7	+	8	=	15
	+		=	
	+		=	
	+		=	
	+		=	

8	+	7	=	15
	+		=	
	+		=	
	+		=	
	+		=	

15	−	7	=	
15	−		=	7

8	+		=	15
	+	8	=	15

15	−	8	=	
	−	8	=	7

8	+	7	=	

Unit 4

47

A

$$3 + 8 = 11$$

$$\square + \square = \square$$

$$\square + \square = \square$$

$$\square + \square = \square$$

$$\square + \square = \square$$

$$8 + 3 = 11$$

$$\square + \square = \square$$

$$\square + \square = \square$$

$$\square + \square = \square$$

$$\square + \square = \square$$

$$11 - 3 = \square$$

$$11 - \square = 8$$

$$3 + \square = 11$$

$$\square + 8 = 11$$

$$11 - 3 = \square$$

$$\square - 8 = 3$$

$$8 + 3 = \square$$

B

$$8 + 4 = 12$$

$$\square + \square = \square$$

$$\square + \square = \square$$

$$\square + \square = \square$$

$$\square + \square = \square$$

$$4 + 8 = 12$$

$$\square + \square = \square$$

$$\square + \square = \square$$

$$\square + \square = \square$$

$$\square + \square = \square$$

$$12 - 8 = \square$$

$$12 - \square = 8$$

$$4 + \square = 12$$

$$\square + 8 = 12$$

$$12 - 8 = \square$$

$$\square - 4 = 8$$

$$4 + 8 = \square$$

C

$$8 + 5 = 13$$

$$\square + \square = \square$$

$$\square + \square = \square$$

$$\square + \square = \square$$

$$\square + \square = \square$$

$$5 + 8 = 13$$

$$\square + \square = \square$$

$$\square + \square = \square$$

$$\square + \square = \square$$

$$\square + \square = \square$$

$$13 - 5 = \square$$

$$13 - \square = 5$$

$$8 + \square = 13$$

$$\square + 5 = 13$$

$$13 - 5 = \square$$

$$\square - 8 = 5$$

$$8 + 5 = \square$$

Repeatedly practising 9 + 2 = 11, 3 + 9 = 12 and 4 + 9 = 13

A

9 + 2 = 11

☐ + ☐ = ☐

☐ + ☐ = ☐

☐ + ☐ = ☐

☐ + ☐ = ☐

2 + 9 = 11

☐ + ☐ = ☐

☐ + ☐ = ☐

☐ + ☐ = ☐

☐ + ☐ = ☐

11 − 2 = ☐

11 − ☐ = 9

9 + ☐ = 11

☐ + 2 = 11

11 − 2 = ☐

☐ − 9 = 2

2 + 9 = ☐

B

3 + 9 = 12

☐ + ☐ = ☐

☐ + ☐ = ☐

☐ + ☐ = ☐

☐ + ☐ = ☐

9 + 3 = 12

☐ + ☐ = ☐

☐ + ☐ = ☐

☐ + ☐ = ☐

☐ + ☐ = ☐

12 − 3 = ☐

12 − ☐ = 3

9 + ☐ = 12

☐ + 3 = 12

12 − 3 = ☐

☐ − 9 = 3

9 + 3 = ☐

C

4 + 9 = 13

☐ + ☐ = ☐

☐ + ☐ = ☐

☐ + ☐ = ☐

☐ + ☐ = ☐

9 + 4 = 13

☐ + ☐ = ☐

☐ + ☐ = ☐

☐ + ☐ = ☐

☐ + ☐ = ☐

13 − 9 = ☐

13 − ☐ = 9

4 + ☐ = 13

☐ + 9 = 13

13 − 9 = ☐

☐ − 4 = 9

4 + 9 = ☐

Unit 4

49

A

5	+	9	=	14	
	+		=		
	+		=		
	+		=		
	+		=		
9	+	5	=	14	
	+		=		
	+		=		
	+		=		
	+		=		
14	–	5	=		
14	–		=	5	
5	+		=	14	
	+	9	=	14	
14	–	9	=		
	–	5	=	9	
5	+	9	=		

B

9	+	6	=	15	
	+		=		
	+		=		
	+		=		
	+		=		
6	+	9	=	15	
	+		=		
	+		=		
	+		=		
	+		=		
15	–	6	=		
15	–		=	6	
9	+		=	15	
	+	6	=	15	
15	–	6	=		
	–	9	=	6	
9	+	6	=		

C

7	+	9	=	16	
	+		=		
	+		=		
	+		=		
	+		=		
9	+	7	=	16	
	+		=		
	+		=		
	+		=		
	+		=		
16	–	7	=		
16	–		=	7	
9	+		=	16	
	+	7	=	16	
16	–	7	=		
	–	9	=	7	
9	+	7	=		

The basic addition facts with answers equal to 100 used to add tens

Ⓐ 4 + 6 = 10 → 4 tens + 6 tens = ☐ tens → 40 + 60 = ☐ hundred

2 + 8 = 10 → 2 tens + 8 tens = ☐ tens → 20 + 80 = ☐ hundred

5 + 5 = 10 → 5 tens + 5 tens = ☐ tens → 50 + 50 = ☐ hundred

3 + 7 = 10 → 3 tens + 7 tens = ☐ tens → 30 + 70 = ☐ hundred

Ⓑ Write a story for 3 tens + 7 tens.

Write your question: ☐

Answer: ☐ ☐

Adding tens with basic facts leads to answers over 100

14 tens equals 1 hundred and 4 tens.

Ⓐ 6 tens + 8 tens = ☐ tens → 6 tens + 8 tens = ☐ hundred + ☐ tens

6 tens + 9 tens = ☐ tens → 6 tens + 9 tens = ☐ hundred + ☐ tens

4 tens + 8 tens = ☐ tens → 4 tens + 8 tens = ☐ hundred + ☐ tens

Ⓑ 1 hundred + 5 tens = ☐ tens

1 hundred + 3 tens = ☐ tens

Ⓒ 1 hundred + 7 tens = ☐ tens

1 hundred + 9 tens = ☐ tens

Unit 4

sixty + eighty = ⬜ hundred + forty

seventy + eighty = ⬜ hundred + ⬜

twenty + ninety = ⬜ hundred + ⬜

eighty + thirty = ⬜ hundred + ⬜

Exercise 62

Three-digit place value

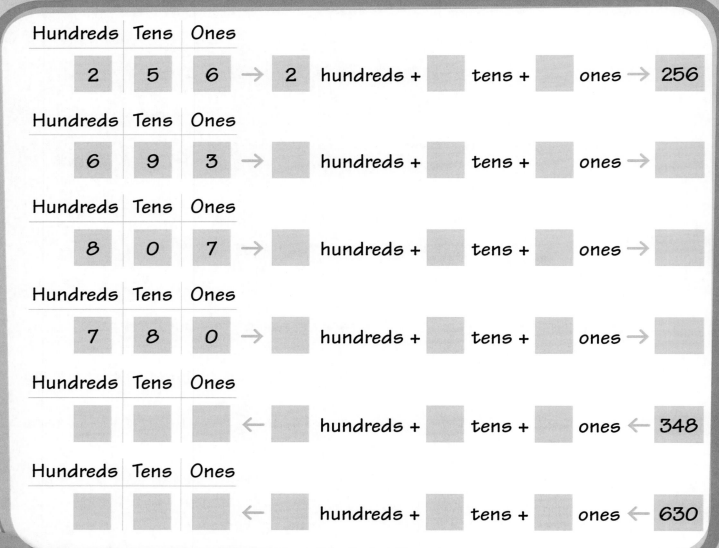

Hundreds	Tens	Ones
2	5	6

→ 2 hundreds + ⬜ tens + ⬜ ones → 256

Hundreds	Tens	Ones
6	9	3

→ ⬜ hundreds + ⬜ tens + ⬜ ones → ⬜

Hundreds	Tens	Ones
8	0	7

→ ⬜ hundreds + ⬜ tens + ⬜ ones → ⬜

Hundreds	Tens	Ones
7	8	0

→ ⬜ hundreds + ⬜ tens + ⬜ ones → ⬜

Hundreds	Tens	Ones
⬜	⬜	⬜

← ⬜ hundreds + ⬜ tens + ⬜ ones ← 348

Hundreds	Tens	Ones
⬜	⬜	⬜

← ⬜ hundreds + ⬜ tens + ⬜ ones ← 630

Writing three-digit place-value numbers in the right order

I write the hundreds before the tens.
Then I write tens before the ones.

A

8 tens + 2 ones + 7 hundreds = 782

7 tens + 9 ones + 5 hundreds =

8 ones + 2 tens + 7 hundreds =

5 ones + 4 hundreds + 3 tens =

B

60 + 400 + 3 = 463

800 + 80 + 4 =

60 + 400 + 9 =

9 + 40 + 300 =

70 + 300 =

7 + 900 =

700 + 1 =

C

200 + 70 + 3 =

7 + 400 + 30 =

6 + 40 + 300 =

90 + 400 + 7 =

80 + 400 =

800 + 4 =

8 + 400 =

D Write a story for $30 + $200 + $50.

Write your question:

Answer: $

Equation: ☐ + ☐ + ☐ = ☐

Unit 4

A

$4 + \boxed{} = 13 \longrightarrow 40 + \boxed{} = 130 \longrightarrow 400 + \boxed{} = 1300$

$6 + \boxed{} = 14 \longrightarrow 60 + \boxed{} = 140 \longrightarrow 600 + \boxed{} = 1400$

$6 + \boxed{} = 13 \longrightarrow 60 + \boxed{} = 130 \longrightarrow 600 + \boxed{} = 1300$

$\boxed{} + 5 = 13 \longrightarrow \boxed{} + 50 = 130 \longrightarrow \boxed{} + 500 = 1300$

$\boxed{} + 6 = 11 \longrightarrow \boxed{} + 60 = 110 \longrightarrow \boxed{} + 600 = 1100$

$\boxed{} + 8 = 17 \longrightarrow \boxed{} + 80 = 170 \longrightarrow \boxed{} + 800 = 1700$

B

$\boxed{}$ drinks + 60 drinks = 140 drinks

50 cakes + $\boxed{}$ cakes = 120 cakes

$\boxed{}$ dollars + 30 dollars = 120 dollars

50 dollars + $\boxed{}$ dollars = 140 dollars

6 thousand dollars + $\boxed{}$ thousand dollars = 14 thousand dollars

8 million dollars + $\boxed{}$ million dollars = 14 million dollars

C Write a story for 60 dollars + some dollars = 110 dollars.

Write your question:

Answer: $\boxed{}$ dollars

Equation: 60 + $\boxed{}$ = 110 dollars

Ⓐ 13 – 9 = ☐ ⟶ 13 tens – 9 tens = ☐ tens ⟶ 130 – 90 = ☐

16 – 7 = ☐ ⟶ 16 tens – 7 tens = ☐ tens ⟶ 160 – 70 = ☐

11 – 3 = ☐ ⟶ 11 tens – 3 tens = ☐ tens ⟶ 110 – 30 = ☐

15 – 6 = ☐ ⟶ 15 tens – 6 tens = ☐ tens ⟶ 150 – 60 = ☐

Ⓑ 1 hundred and forty – 80 = ☐ ⟶ 140 – 80 = ☐

1 hundred and thirty – 60 = ☐ ⟶ 130 – 60 = ☐

170 – ninety = ☐ ⟶ 170 – 90 = ☐

110 – twenty = ☐ ⟶ 110 – 20 = ☐

160 – eighty = ☐ ⟶ 160 – 80 = ☐

Ⓒ Write a story for 120 – eighty.

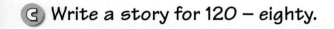

Write your question: ☐

Write the answer: ☐

Write the equation: 120 – 80 = ☐

Unit 4

A

$13 - \boxed{} = 4 \longrightarrow 130 - \boxed{} = 40 \longrightarrow 1300 - \boxed{} = 400$

$12 - \boxed{} = 6 \longrightarrow 120 - \boxed{} = 60 \longrightarrow 1200 - \boxed{} = 600$

$13 - \boxed{} = 6 \longrightarrow 130 - \boxed{} = 60 \longrightarrow 1300 - \boxed{} = 600$

$\boxed{} - 5 = 8 \longrightarrow \boxed{} - 50 = 80 \longrightarrow \boxed{} - 500 = 800$

$\boxed{} - 7 = 5 \longrightarrow \boxed{} - 70 = 50 \longrightarrow \boxed{} - 700 = 500$

$\boxed{} - 8 = 9 \longrightarrow \boxed{} - 80 = 90 \longrightarrow \boxed{} - 800 = 900$

B 13 thousand bananas − $\boxed{}$ thousand bananas = 9 thousand bananas

$\boxed{}$ dollars − 70 dollars = 60 dollars

1300 dollars − $\boxed{}$ dollars = 900 dollars

$\boxed{}$ DVDs + 6 DVDs + 8 DVDs = 17 $\boxed{}$

$\$3 + \$9 - \$\boxed{} = \$11$

C Write a story for \$140 − some money = \$40.

[blank lines]

Write your question: [blank]

[blank lines]

Write the answer: $\boxed{}$ dollars

Write the equation: $140 - \boxed{} = 40$

Patterns in forwards and backwards counting by tens

100	110	120	130						
200	210		230		250				290

350	360	370				410		430	440
450		470	480	490		510		530	

255	265	275		295		315		335	345
355		375	385	395			425		

50	60	70	80		100		120		140
150		170		190		210		230	
250	260						320		340

Unit 4

Doubles plus one or two

I need to know 2 + 2 = 4, 6 + 6 = 12, 3 + 3 = 6, 7 + 7 = 14,
4 + 4 = 8, 8 + 8 = 16, 5 + 5 = 10, 9 + 9 = 18

4 + 5 = ☐	←	4 + 4 = ☐	→	5 + 4 = ☐
4 + 3 = ☐	←	3 + 3 = ☐	→	3 + 4 = ☐
7 + 8 = ☐	←	7 + 7 = ☐	→	8 + 7 = ☐
6 + 7 = ☐	←	6 + 6 = ☐	→	7 + 6 = ☐
6 + 8 = ☐	←	6 + 6 = ☐	→	8 + 6 = ☐
5 + 7 = ☐	←	5 + 5 = ☐	→	7 + 5 = ☐

Adding tens using doubles and near doubles facts

(A) 4 + 4 = 8 → 4 tens + 4 tens = ☐ tens → 40 + 40 = ☐

2 + 3 = 5 → 2 tens + 3 tens = ☐ tens → 20 + 30 = ☐

5 + 4 = 9 → 5 tens + 4 tens = ☐ tens → 50 + 40 = ☐

3 + 5 = 8 → 3 tens + 5 tens = ☐ tens → 30 + 50 = ☐

4 + 2 = 6 → 4 tens + 2 tens = ☐ tens → 40 + 20 = ☐

(B) John has five ten dollar notes, and he gets another three ten dollar
notes for his birthday. How much money does he have now?

Write the answer: ☐ dollars

Addition facts lead to subtraction facts

> I know 4 + 5 equals 9. So 9 minus 4 equals 5, and 9 minus 5 equals 4.

9 − 5 = ☐	←	4 + 5 = ☐	→	9 − 4 = ☐
7 − 3 = ☐	←	4 + 3 = ☐	→	7 − 4 = ☐
13 − 7 = ☐	←	7 + 6 = ☐	→	13 − 6 = ☐
11 − 6 = ☐	←	6 + 5 = ☐	→	11 − 5 = ☐

Using addition facts to add three numbers

A
5 + 5 + 8 = ☐

8 + 8 + 3 = ☐

8 + 7 + 3 = ☐

6 + 7 + 6 = ☐

B
9 + 8 + 2 = ☐

7 + 6 + 4 = ☐

6 + 6 + 7 = ☐

3 + 3 + 10 = ☐

C
7 + 9 + 2 = ☐

5 + 6 + 7 = ☐

5 + 5 + 9 = ☐

4 + 4 + 10 = ☐

D Write a story for $150 − some money = $100.

Write your question: ☐

Write the equation: 150 − ☐ = 100

Unit 5

Extras: using addition facts that add up to ten

$$5 + 4 + 8 + 5 + 2 + 6 + 9 = \boxed{39}$$

Ⓐ 5 + 4 + 8 + 5 + 2 + 6 + 9 = ☐

1 + 4 + 9 + 5 + 8 + 6 + 5 = ☐

6 + 4 + 9 + 5 + 8 + 1 + 5 + 2 = ☐

Ⓑ 8 + 4 + 6 + 5 + 2 + 5 + 8 = ☐

2 + 2 + 8 + 8 + 6 + 4 + 3 = ☐

2 + 8 + 8 + 3 + 7 + 5 + 5 + 2 = ☐

Using subtraction facts to subtract with three numbers

Ⓐ 12 − 6 − 3 = ☐

14 − 7 − 6 = ☐

18 − 9 − 2 = ☐

19 − 10 − 4 = ☐

Ⓑ 18 − 9 − 4 = ☐

18 − 9 − 7 = ☐

16 − 8 − 2 = ☐

17 − 7 − 5 = ☐

Ⓒ 16 − 8 − 6 = ☐

12 − 6 − 2 = ☐

14 − 7 − 4 = ☐

15 − 7 − 4 = ☐

Ⓓ Write a story for 19 − 9 − 5.

Write your question: ☐

Write the answer: ☐

Write the equation: 19 − 9 − 5 = ☐

Repeatedly practising 2 + 2 = 4, 3 + 3 = 6 and 5 + 5 =10

A

2 + 2 = 4

☐ + ☐ = ☐

☐ + ☐ = ☐

☐ + ☐ = ☐

☐ + ☐ = ☐

☐ + ☐ = ☐

☐ + ☐ = ☐

☐ + ☐ = ☐

☐ + ☐ = ☐

☐ + ☐ = ☐

4 – 2 = ☐

4 – ☐ = 2

2 + ☐ = 4

☐ + 2 = 4

4 – 2 = ☐

☐ – 2 = 2

2 + 2 = ☐

B

3 + 3 = 6

☐ + ☐ = ☐

☐ + ☐ = ☐

☐ + ☐ = ☐

☐ + ☐ = ☐

☐ + ☐ = ☐

☐ + ☐ = ☐

☐ + ☐ = ☐

☐ + ☐ = ☐

☐ + ☐ = ☐

6 – 3 = ☐

6 – ☐ = 3

3 + ☐ = 6

☐ + 3 = 6

6 – 3 = ☐

☐ – 3 = 3

3 + 3 = ☐

C

5 + 5 = 10

☐ + ☐ = ☐

☐ + ☐ = ☐

☐ + ☐ = ☐

☐ + ☐ = ☐

☐ + ☐ = ☐

☐ + ☐ = ☐

☐ + ☐ = ☐

☐ + ☐ = ☐

☐ + ☐ = ☐

10 – 5 = ☐

10 – ☐ = 5

5 + ☐ = 10

☐ + 5 = 10

10 – 5 = ☐

☐ – 5 = 5

5 + 5 = ☐

A

5 + 4 = 9

□ + □ = □

□ + □ = □

□ + □ = □

□ + □ = □

4 + 5 = 9

□ + □ = □

□ + □ = □

□ + □ = □

□ + □ = □

9 − 4 = □

9 − □ = 4

5 + □ = 9

□ + 5 = 9

9 − 4 = □

□ − 5 = 4

5 + 4 = □

B

6 + 6 = 12

□ + □ = □

□ + □ = □

□ + □ = □

□ + □ = □

□ + □ = □

□ + □ = □

□ + □ = □

□ + □ = □

□ + □ = □

12 − 6 = □

12 − □ = 6

6 + □ = 12

□ + 6 = 12

12 − 6 = □

□ − 6 = 6

6 + 6 = □

C

5 + 6 = 11

□ + □ = □

□ + □ = □

□ + □ = □

□ + □ = □

6 + 5 = 11

□ + □ = □

□ + □ = □

□ + □ = □

□ + □ = □

11 − 6 = □

11 − □ = 6

5 + □ = 11

□ + 5 = 11

11 − 5 = □

□ − 6 = 5

5 + 6 = □

Repeatedly practising 7 + 6 = 13, 7 + 7 = 14 and 7 + 8 = 15

A

7 + 6 = 13

☐ + ☐ = ☐

☐ + ☐ = ☐

☐ + ☐ = ☐

☐ + ☐ = ☐

6 + 7 = 13

☐ + ☐ = ☐

☐ + ☐ = ☐

☐ + ☐ = ☐

☐ + ☐ = ☐

13 – 7 = ☐

13 – ☐ = 7

6 + ☐ = 13

☐ + 7 = 13

13 – 6 = ☐

☐ – 6 = 7

13 + 6 = ☐

B

7 + 7 = 14

☐ + ☐ = ☐

☐ + ☐ = ☐

☐ + ☐ = ☐

☐ + ☐ = ☐

☐ + ☐ = ☐

☐ + ☐ = ☐

☐ + ☐ = ☐

☐ + ☐ = ☐

☐ + ☐ = ☐

14 – 7 = ☐

14 – ☐ = 7

7 + ☐ = 14

☐ + 7 = 14

14 – 7 = ☐

☐ – 7 = 7

7 + 7 = ☐

C

7 + 8 = 15

☐ + ☐ = ☐

☐ + ☐ = ☐

☐ + ☐ = ☐

☐ + ☐ = ☐

8 + 7 = 15

☐ + ☐ = ☐

☐ + ☐ = ☐

☐ + ☐ = ☐

☐ + ☐ = ☐

15 – 7 = ☐

15 – ☐ = 7

8 + ☐ = 15

☐ + 8 = 15

15 – 8 = ☐

☐ – 7 = 8

7 + 8 = ☐

Repeatedly practising 8 + 8 = 16, 9 + 8 = 17 and 9 + 9 = 18

A

$8 + 8 = 16$

$\square + \square = \square$

$\square + \square = \square$

$\square + \square = \square$

$\square + \square = \square$

$\square + \square = \square$

$\square + \square = \square$

$\square + \square = \square$

$\square + \square = \square$

$\square + \square = \square$

$16 - 8 = \square$

$16 - \square = 8$

$8 + \square = 16$

$\square + 8 = 16$

$16 - 8 = \square$

$\square - 8 = 8$

$8 + 8 = \square$

B

$9 + 8 = 17$

$\square + \square = \square$

$\square + \square = \square$

$\square + \square = \square$

$\square + \square = \square$

$8 + 9 = 17$

$\square + \square = \square$

$\square + \square = \square$

$\square + \square = \square$

$17 - 8 = \square$

$17 - \square = 9$

$8 + \square = 17$

$\square + 8 = 17$

$17 - 9 = \square$

$\square - 9 = 8$

$9 + 8 = \square$

C

$9 + 9 = 18$

$\square + \square = \square$

$\square + \square = \square$

$\square + \square = \square$

$\square + \square = \square$

$\square + \square = \square$

$\square + \square = \square$

$\square + \square = \square$

$\square + \square = \square$

$18 - 9 = \square$

$18 - \square = 9$

$9 + \square = 18$

$\square + 9 = 18$

$18 - 9 = \square$

$\square - 9 = 9$

$9 + 9 = \square$

Adding tens using words like 5 tens and fifty

> 4 plus 3 equals 7.
> Forty plus thirty equals seventy.

5 tens + 4 tens = ⬜ tens   ⟶   50 + 40 = ⬜

2 tens + 3 tens = ⬜ tens   ⟶   20 + 30 = ⬜

50 + 3 tens = ⬜ tens   ⟶   50 + 30 = ⬜

forty + 50 = ⬜ tens   ⟶   40 + 50 = ⬜

4 tens + thirty = ⬜ tens   ⟶   40 + 30 = ⬜

Adding three numbers with tens

A   5 tens + 3 tens + 2 tens ⟶ ⬜ tens ⟶ ⬜ hundred

4 tens + 4 tens + 2 tens ⟶ ⬜ tens ⟶ ⬜ hundred

3 tens + 2 tens + 5 tens ⟶ ⬜ tens ⟶ ⬜ hundred

3 tens + 2 tens + ⬜ tens ⟵ 10 tens ⟶ ⬜ hundred

4 tens + ⬜ tens + 2 tens ⟵ 10 tens ⟶ ⬜ hundred

3 tens + 2 tens + ⬜ tens ⟵ ⬜ tens ⟵ 1 hundred

B   Jane has 2 ten dollar notes. Bob has 5 ten dollar notes and Freda has 3 ten dollar notes. How much money do they have altogether?

Answer: ⬜ dollars

Unit 5

Adding two-digit numbers like 30 and 45

I group my tens together first.

30 + 26 ⟶	3	tens +	2	tens +	ones ⟶	56
60 + 27 ⟶		tens +		tens +	ones ⟶	
20 + 66 ⟶		tens +		tens +	ones ⟶	
20 + 69 ⟶		tens +		tens +	ones ⟶	
50 + 48 ⟶		tens +		tens +	ones ⟶	

Adding two-digit numbers like 34 and 45

Ⓐ 33 + 26 ⟶ 5 tens + 9 ones ⟶ 33 + 26 = 59

65 + 22 ⟶ ☐ tens + ☐ ones ⟶ 65 + 22 = ☐

12 + 76 ⟶ ☐ tens + ☐ ones ⟶ 12 + 76 = ☐

36 + 50 ⟶ ☐ tens + ☐ ones ⟶ 36 + 50 = ☐

72 + 27 ⟶ ☐ tens + ☐ ones ⟶ 72 + 27 = ☐

Ⓑ Mica buys an MP3 music player for $74 and a CD for $13. How much does Mica spend altogether?

Write the answer: $ ☐

Write the equation: ☐ + ☐ = ☐

Adding tens with answers over 100

I use my basic facts then replace
ten tens by a hundred.

A

70 + 80 →	15	tens →	1	hundred +		tens →	150
60 + 70 →		tens →		hundred +		tens →	
90 + 90 →		tens →		hundred +		tens →	
90 + 80 →		tens →		hundred +		tens →	
70 + 50 →		tens →		hundred +		tens →	

B

40 + 80 =

70 + 90 =

60 + 80 =

50 + 80 =

C

90 + 60 =

30 + 80 =

20 + 90 =

90 + 50 =

D

sixty + ninety = 150

sixty + seventy =

seventy + fifty =

sixty + fifty =

eighty + thirty =

fifty + sixty =

E

7 tens + eighty = 150

9 tens + twenty =

60 + seven tens =

90 + seven tens =

40 + eight tens =

80 + eight tens =

Adding hundreds using basic facts

2 plus 5 is 7,
so 2 hundreds + 5 hundreds = 7 hundreds.

A 4 hundreds + 3 hundreds = ☐ hundreds ⟶ 400 + 300 = ☐

5 hundreds + 4 hundreds = ☐ hundreds ⟶ 500 + 400 = ☐

4 hundreds + 4 hundreds = ☐ hundreds ⟶ 400 + 400 = ☐

2 hundreds + 3 hundreds = ☐ hundreds ⟶ 200 + 300 = ☐

B

400 + 500 = ☐

400 + 300 = ☐

800 + 100 = ☐

C

300 + 400 = ☐

500 + 300 = ☐

100 + 700 = ☐

D

200 + 3 hundreds = ☐

1 hundred + 400 = ☐

7 hundreds + 200 = ☐

E 100 + 5 hundreds = ☐

4 hundreds + 400 = ☐

5 hundreds + 200 = ☐

F Geraldine buys a TV for $300 and a games player for four hundred dollars. How much does she pay the shopkeeper?

Answer: $ ☐

G The Jones family has monthly power bills for $100, $200 and three hundred dollars. How much does the family pay for its power over the three months?

Answer: $ ☐ apples

Unit 5

I always replace ten hundreds
with one thousand.

$7 + 6 = 13$

7 hundreds + 6 hundreds = 13 hundreds

13 hundreds = 1 thousand + 3 hundreds

A 5 hundreds + 6 hundreds = ☐ thousand + ☐ hundreds

7 hundreds + 6 hundreds = ☐ thousand + ☐ hundreds

8 hundreds + 800 = ☐ thousand + ☐ hundreds

700 + 700 = ☐ thousand + ☐ hundreds

1 thousand + 4 hundreds + 0 tens + 0 ones = 1400

B

900 + 800 = 1700

700 + 800 = ☐

800 + 800 = ☐

500 + 600 = ☐

C

800 + 300 + 100 = ☐

900 + 200 + 100 = ☐

700 + 100 + 700 = ☐

600 + 600 + 200 = ☐

D Briony has three bank accounts. The amounts she has in the accounts are $400, $400 and $500. How much money does Briony have altogether?

Answer: $ ☐

E Neil sells 700 apples at the market on Saturday and he sells 600 apples on Sunday. How many apples does he sell at the market in the weekend?

Answer: ☐

Ⓐ 4 + ☐ = 9 ⟶ 40 + ☐ = 90 ⟶ 400 + ☐ = 900

6 + ☐ = 12 ⟶ 60 + ☐ = 120 ⟶ 600 + ☐ = 1200

6 + ☐ = 13 ⟶ 60 + ☐ = 130 ⟶ 600 + ☐ = 1300

☐ + 5 = 9 ⟶ ☐ + 50 = 90 ⟶ ☐ + 500 = 900

☐ + 6 = 11 ⟶ ☐ + 60 = 110 ⟶ ☐ + 600 = 1100

☐ + 9 = 18 ⟶ ☐ + 90 = 180 ⟶ ☐ + 900 = 1800

Ⓑ ☐ drinks + 6 drinks = 12 drinks

5 cakes + ☐ cakes = 10 cakes

30 ☐ + 30 dollars = 60 dollars

500 dollars + ☐ = 900 dollars

3000 dollars + ☐ = 7000 dollars

Ⓒ Write a story for 400 + a number equals 9 hundred.

Write your question:

Write the answer:

Write the equation: 400 + ☐ = 900

Subtracting tens using basic facts

I know 16 minus 8 equals 8.
So 160 minus 80 equals 80.

A 7 − 3 = 4 ⟶ 7 tens − 3 tens = ☐ tens ⟶ 70 − 30 = ☐

9 − 5 = 4 ⟶ 9 tens − 5 tens = ☐ tens ⟶ 90 − 50 = ☐

6 − 3 = 3 ⟶ 6 tens − 3 tens = ☐ tens ⟶ 60 − 30 = ☐

7 − 4 = 3 ⟶ 7 tens − 4 tens = ☐ tens ⟶ 70 − 40 = ☐

8 − 4 = 4 ⟶ 8 tens − 4 tens = ☐ tens ⟶ 80 − 40 = ☐

9 − 4 = 5 ⟶ 9 tens − 4 tens = ☐ tens ⟶ 90 − 40 = ☐

B

160 − 80 = ☐

160 − 70 = ☐

1 hundred and forty − 80 = ☐

1 hundred and eighty − 90 = ☐

C

90 − 40 = ☐

170 − 90 = ☐

130 − sixty = ☐

110 − sixty = ☐

D Write a story for 150 − 8 tens.

Write your question: ☐

Write the answer: ☐

Write the equation: 150 − 80 = ☐

Extras: more practice applying the basic subtraction facts

A 9 − ☐ = 4 ⟶ 90 − ☐ = 40 ⟶ 900 − ☐ = 400

12 − ☐ = 6 ⟶ 120 − ☐ = 60 ⟶ 1200 − ☐ = 600

13 − ☐ = 6 ⟶ 130 − ☐ = 60 ⟶ 1300 − ☐ = 600

☐ − 5 = 4 ⟶ ☐ − 50 = 40 ⟶ ☐ − 500 = 400

☐ − 6 = 5 ⟶ ☐ − 60 = 50 ⟶ ☐ − 600 = 500

☐ − 9 = 9 ⟶ ☐ − 90 = 90 ⟶ ☐ − 900 = 900

B 17 thousand bananas − ☐ thousand bananas = 9 thousand bananas

☐ dollars − 30 dollars = 40 dollars

500 dollars − ☐ dollars = 300 dollars

9000 dollars − 5000 ☐ = 4000 dollars

☐ DVDs − 60 DVDs = 70 ☐

C Write a story for 900 minus a number equals 500.

Write your question:

Write the answer: ☐

Write the equation: 900 − ☐ = 500

Skip counting forwards and backwards by twos

2	4	6	8				16		20
22		26		30					40

70	68	66		62		58			52
50		46		42					32

	302	304		308	310				318
320		324				332		336	

	618	616		612			606		
	598		594			588			582
					570		566		562

A

4 + 4 =		
5 + 6 =		
7 + 7 =		
2 + 7 =		
4 + ___ = 8		
6 + ___ = 13		
9 + ___ = 18		
4 + ___ = 6		
___ + 6 = 12		
___ + 2 = 4		
___ + 7 = 12		
___ + 6 = 12		
13 − 6 =		
18 − 9 =		
8 − 4 =		
9 − 8 =		
30 + 50 =		
30 + 60 =		
70 + 70 =		
70 + 80 =		

Score: ___

B

6 + 6 =		
7 + 5 =		
8 + 8 =		
6 + 2 =		
5 + ___ = 10		
7 + ___ = 12		
7 + ___ = 14		
2 + ___ = 7		
___ + 4 = 8		
___ + 1 = 2		
___ + 6 = 13		
___ + 5 = 10		
11 − 7 =		
16 − 8 =		
6 − 3 =		
7 − 6 =		
50 + 60 =		
50 + 40 =		
60 + 60 =		
70 + 60 =		

Score: ___

C

2 + 2 =		
3 + 8 =		
9 + 9 =		
8 + 2 =		
3 + ___ = 6		
5 + ___ = 13		
6 + ___ = 12		
2 + ___ = 9		
___ + 5 = 10		
___ + 3 = 6		
___ + 8 = 15		
___ + 9 = 18		
12 − 7 =		
14 − 8 =		
4 − 2 =		
8 − 7 =		
80 + 70 =		
40 + 30 =		
90 + 90 =		
40 + 60 =		

Score: ___

Unit 5

D

7 + 7 =

5 + 6 =

8 + 9 =

3 + 3 =

9 +   = 17

4 +   = 11

5 +   = 12

3 +   = 6

   + 5 = 10

   + 7 = 13

   + 4 = 8

   + 4 = 13

11 – 7 =

11 – 6 =

6 – 4 =

18 – 9 =

20 + 40 =

60 + 50 =

80 + 60 =

90 + 90 =

Score:

E

8 + 8 =

7 + 5 =

7 + 8 =

2 + 2 =

9 +   = 16

6 +   = 11

5 +   = 13

4 +   = 8

   + 8 = 16

   + 6 = 13

   + 3 = 6

   + 3 = 12

16 – 8 =

13 – 6 =

7 – 6 =

17 – 9 =

30 + 40 =

70 + 60 =

90 + 70 =

60 + 70 =

Score:

F

9 + 9 =

6 + 7 =

9 + 7 =

5 + 5 =

9 +   = 15

6 +   = 13

5 +   = 14

2 +   = 4

   + 7 = 14

   + 9 = 17

   + 2 = 4

   + 6 = 15

12 – 6 =

11 – 5 =

9 – 8 =

17 – 8 =

50 + 20 =

60 + 70 =

60 + 80 =

70 + 80 =

Score:

Unit 5

75

G

6	+	6	=	
5	+	7	=	
8	+	9	=	
4	+	4	=	
6	+		=	14
7	+		=	12
5	+		=	6
3	+		=	10
	+	7	=	12
	+	5	=	7
	+	8	=	16
	+	1	=	7
17	–	8	=	
10	–	8	=	
14	–	6	=	
12	–	9	=	
20	+	20	=	
60	+	50	=	
80	+	70	=	
90	+	80	=	

Score:

H

9	+	9	=	
4	+	7	=	
7	+	9	=	
5	+	5	=	
7	+		=	15
7	+		=	13
8	+		=	9
2	+		=	4
	+	6	=	11
	+	8	=	10
	+	7	=	14
	+	2	=	6
14	–	8	=	
10	–	6	=	
16	–	9	=	
15	–	8	=	
30	+	20	=	
70	+	50	=	
70	+	90	=	
60	+	70	=	

Score:

I

5	+	5	=	
4	+	8	=	
9	+	6	=	
3	+	3	=	
8	+		=	15
8	+		=	12
1	+		=	7
4	+		=	8
	+	8	=	13
	+	2	=	11
	+	9	=	18
	+	1	=	8
15	–	7	=	
10	–	4	=	
15	–	8	=	
11	–	9	=	
30	+	40	=	
50	+	60	=	
50	+	70	=	
70	+	80	=	

Score:

Speed test graph

After each one-minute test, mark the number you got correct with a line. Colour in the bar below your line. Then write the date.

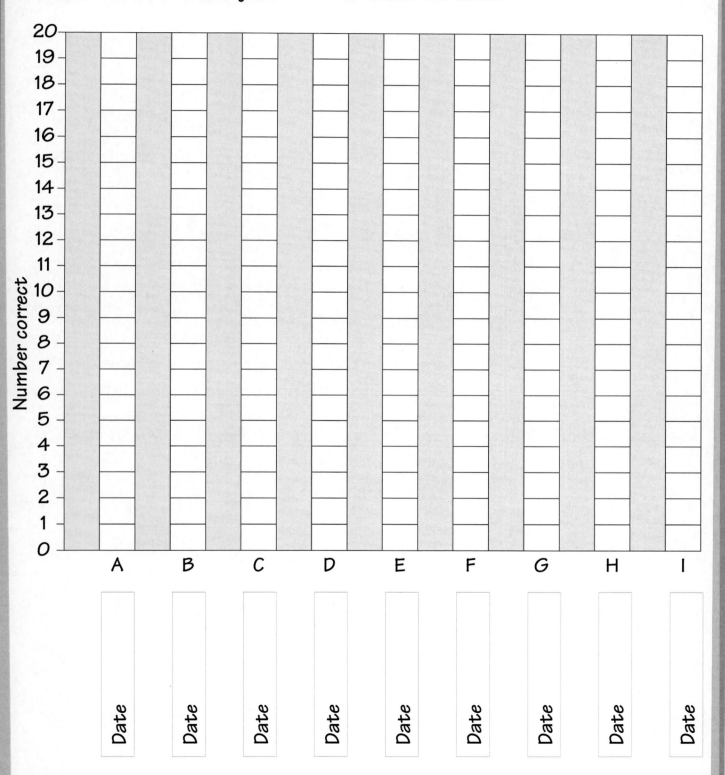

Number correct

20
19
18
17
16
15
14
13
12
11
10
9
8
7
6
5
4
3
2
1
0

A   B   C   D   E   F   G   H   I

Date   Date   Date   Date   Date   Date   Date   Date   Date

Unit 5

77

Adding or subtracting zero

I know that adding or subtracting zero does not change the number.

$8 + 0 = \boxed{}$ $0 + 8 = \boxed{}$ $8 - 0 = \boxed{}$

$9 + 0 = \boxed{}$ $0 + 9 = \boxed{}$ $9 - 0 = \boxed{}$

$24 + 0 = \boxed{}$ $0 + 24 = \boxed{}$ $24 - 0 = \boxed{}$

$8870 + 0 = \boxed{}$ $0 + 8870 = \boxed{}$ $8870 - 0 = \boxed{}$

One answer: many facts

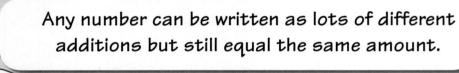

Any number can be written as lots of different additions but still equal the same amount.

$9 = 8 + 1$ and $9 = 7 + \boxed{}$ and $9 = \boxed{} + 6$

$12 = 8 + 4$ and $12 = 5 + \boxed{}$ and $12 = \boxed{} + 6$

$10 = 3 + \boxed{}$ and $10 = 4 + \boxed{}$ and $10 = 5 + \boxed{}$

$7 = 4 + \boxed{}$ and $7 = 1 + \boxed{}$ and $7 = \boxed{} + 5$

$13 = 9 + \boxed{}$ and $13 = \boxed{} + 6$ and $13 = \boxed{} + 8$

$8 = 6 + 2$ and $8 = \boxed{} + 4$ and $8 = \boxed{} + 5$

Swapping tens and ones

When subtracting I know sometimes I must swap one ten for ten ones.

A
1 ten + 6 ones = 16 ones

1 ten + 2 ones = ___ ones

2 tens + 6 ones = ___ ones

1 ten + ___ ones = 13 ones

B
1 ten + 9 ones = ___ ones

1 ten + 0 ones = ___ ones

1 ten + ___ ones = 19 ones

___ tens + 2 ones = 42 ones

C
3 tens + 12 ones = ⟶ 4 tens + 2 ones = 42

4 tens + 18 ones = ⟶ 5 tens + ___ ones = ___

3 tens + 17 ones = ⟶ ___ tens + ___ ones = 47

___ tens + 16 ones = ⟵ ___ tens + 6 ones = 76

D
5 tens + 13 ones = 63

2 tens + 18 ones = ___

8 tens + 19 ones = ___

E
7 tens + 25 ones = 95

3 tens + 24 ones = ___

4 tens + 23 ones = ___

F John has 3 ten-dollar notes and 5 one-dollar coins. Mary has 2 ten-dollar notes and 15 one-dollar coins. Who has more money, John or Mary? Or do they both have the same amount?

Answer: _____

A

3 tens + 5 ones are [more] than 2 tens + 14 ones.

5 tens + 3 ones are [   ] than 4 tens + 11 ones.

6 tens + 5 ones are [less] than 5 tens + 18 ones.

8 tens + 3 ones are [   ] than 7 tens + 12 ones.

3 tens + 15 ones are [   ] than 4 tens + 4 ones.

3 tens + 16 ones are [   ] than 4 tens + 2 ones.

2 tens + 14 ones are [   ] to 3 tens + 4 ones.

I know < means *less than*. I know > means *more than*, and = means *equal to*.

B

2 tens + 13 ones [<] 3 tens + 4 ones.

3 tens + 8 ones [   ] 2 tens + 19 ones.

7 tens + 4 ones [>] 6 tens + 13 ones.

4 tens + 12 ones [   ] 3 tens + 10 ones.

7 tens + 4 ones [   ] 6 tens + 13 ones.

5 tens + 17 ones [   ] 6 tens + 7 ones.

C 70 + 5 [>] 60 + 12

40 + 4 [   ] 30 + 12

50 + 5 [   ] 40 + 14

D 10 + 19 [   ] 20 + 8

50 + 13 [   ] 60 + 3

90 + 4 [   ] 80 + 14

Extras: adding two-digit numbers like 28 and 57

When adding, I know I must swap ten ones for one ten.

38 → **3** tens + **8** ones

+ 26 → **2** tens + **6** ones

5 tens + **14** ones = **64**

---

47 → **4** tens + ☐ ones

+ 38 → ☐ tens + ☐ ones

☐ tens + **15** ones = ☐

---

17 → ☐ tens + ☐ ones

+ 38 → ☐ tens + ☐ ones

4 tens + ☐ ones = ☐

---

57 → ☐ tens + ☐ ones

+ 19 → ☐ tens + ☐ ones

☐ tens + ☐ ones = ☐

---

69 → ☐ tens + ☐ ones

+ 25 → ☐ tens + ☐ ones

☐ tens + ☐ ones = ☐

Extras: adding vertically with answers under 100

I say to myself: 5 ones + 9 ones = 14 ones, then 14 ones = 1 ten + 4 ones. I write 4 in the ones column. I write 1 in the tens column. Then I say to myself, 1 ten + 6 tens + 2 tens = 9 tens. I write 9 in the tens column.

```
    1
   65
 + 29
 ----
   94
```

```
   45          65          64          25
 + 26        + 18        + 26        + 47
 ----        ----        ----        ----
```

```
   33          50          55          45
 + 58        + 28        + 39        + 25
 ----        ----        ----        ----
```

```
   65          79          47          13
 + 22        + 19        + 36        + 66
 ----        ----        ----        ----
```

```
   42          44          43          15
 + 28        + 44        + 26        + 77
 ----        ----        ----        ----
```

Practising the basic addition facts and problem solving

+	2	7	5		6	0	3		1	4
7				16				15		
6										
9										
		12						13		

More practising the basic addition facts and problem solving

A

+	4	6		9
7				
			12	14
9			16	
6				

B

+			5	9	3
4					
		14			
5		11			
		13			

C

+	40	60		90		50	20	30
70			80					
				140				
90					160		110	

Subtracting two-digit numbers like 45 – 30

76 – 20 = [7] tens + [6] ones – [2] tens = [5] tens + [6] ones = [56]

98 – 60 = [9] tens + [8] ones – [6] tens = [ ] tens + [8] ones = [ ]

51 – 30 = [5] tens + [ ] ones – [3] tens = [ ] tens + [ ] ones = [ ]

67 – 50 = [ ] tens + [ ] ones – [ ] tens = [ ] tens + [ ] ones = [ ]

83 – 40 = [ ] tens + [ ] ones – [ ] tens = [ ] tens + [ ] ones = [ ]

Subtracting two-digit numbers like 48 – 26

48 → [4] tens + [8] ones
– 26 → [2] tens + [6] ones
[2] tens + [2] ones = [22]

98 → [9] tens + [ ] ones
– 26 → [ ] tens + [6] ones
[7] tens + [ ] ones = [ ]

45 → [4] tens + [ ] ones
– 14 → [ ] tens + [ ] ones
[ ] tens + [ ] ones = [31]

74 → **6** tens + **14** ones
− 26 → **2** tens + **6** ones
4 tens + **8** ones = **48**

81 → **7** tens + **11** ones
− 38 → ☐ tens + ☐ ones
☐ tens + ☐ ones = ☐

42 → ☐ tens + **12** ones
− 14 → ☐ tens + ☐ ones
☐ tens + ☐ ones = **28**

62 → ☐ tens + **12** ones
− 33 → ☐ tens + **3** ones
☐ tens + ☐ ones = ☐

73 → ☐ tens + ☐ ones
− 55 → ☐ tens + ☐ ones
☐ tens + ☐ ones = ☐

62 → ☐ tens + ☐ ones
− 59 → ☐ tens + ☐ ones
☐ tens + ☐ ones = ☐

Two times

8 + 8 = 16 $\longrightarrow$ 2 lots of 8 = ☐ $\longrightarrow$ 2 x 8 = ☐

6 + 6 = ☐ $\longrightarrow$ 2 lots of 6 = ☐ $\longrightarrow$ 2 x 6 = ☐

5 + 5 = ☐ $\longrightarrow$ 2 lots of 5 = ☐ $\longrightarrow$ 2 x 5 = ☐

40 + 40 = ☐ $\longrightarrow$ 2 lots of 40 = ☐ $\longrightarrow$ 2 x 40 = ☐

Times ten

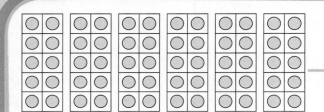 $\longrightarrow$ 6 lots of 10 $\longrightarrow$ 6 x 10 = ☐

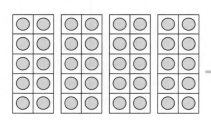 $\longrightarrow$ ☐ lots of 10 $\longrightarrow$ 4 x ☐ = ☐

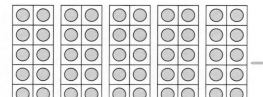 $\longrightarrow$ 5 lots of ☐ $\longrightarrow$ ☐ x ☐ = ☐

I know how to skip count
to do multiplication.

A

$2 + 2 + 2 + 2 + 2 + 2 + 2 = \boxed{}$ ⟵ $7 \times 2 = 14$

$2 + 2 + 2 + 2 + 2 + 2 = \boxed{}$ ⟷ $6 \times \boxed{} = \boxed{}$

$2 + 2 + 2 + 2 + 2 = \boxed{}$ ⟷ $\boxed{} \times 2 = \boxed{}$

$5 + 5 + 5 + 5 + 5 + 5 + 5 = \boxed{}$ ⟷ $7 \times 5 = \boxed{}$

$5 + 5 + 5 + 5 + 5 + 5 = \boxed{}$ ⟷ $6 \times \boxed{} = \boxed{}$

B

$\boxed{} + \boxed{} + \boxed{} + \boxed{} + \boxed{} + \boxed{} = \boxed{}$ ⟵ $6 \times 10 = 60$

$\boxed{} + \boxed{} + \boxed{} + \boxed{} + \boxed{} = \boxed{}$ ⟷ $5 \times 10 = \boxed{}$

$\boxed{} + \boxed{} + \boxed{} + \boxed{} + \boxed{} = \boxed{}$ ⟷ $5 \times 5 = \boxed{}$

$\boxed{} + \boxed{} + \boxed{} + \boxed{} = \boxed{}$ ⟷ $4 \times 10 = \boxed{}$

C Five children each have $2. How much money do they have altogether?

Answer: $ \boxed{}

Write the equation: $\boxed{} \times \boxed{} = \boxed{}$

The five times facts by regrouping fives into tens

I use the fact that five plus five equals ten.

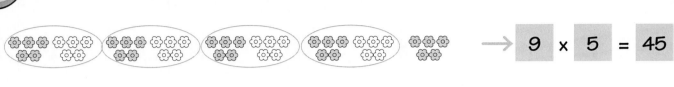

→ 9 x 5 = 45

→ 6 x 5 = ☐

→ ☐ x 5 = ☐

Multiplication as an array

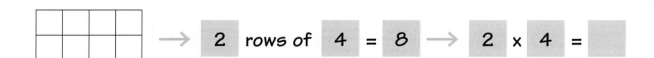

→ 2 rows of 4 = 8 → 2 x 4 = ☐

→ 2 rows of 5 = ☐ → 2 x ☐ = ☐

→ 3 rows of ☐ = ☐ → 3 x ☐ = ☐

→ 4 rows of ☐ = ☐ → ☐ x ☐ = ☐

A

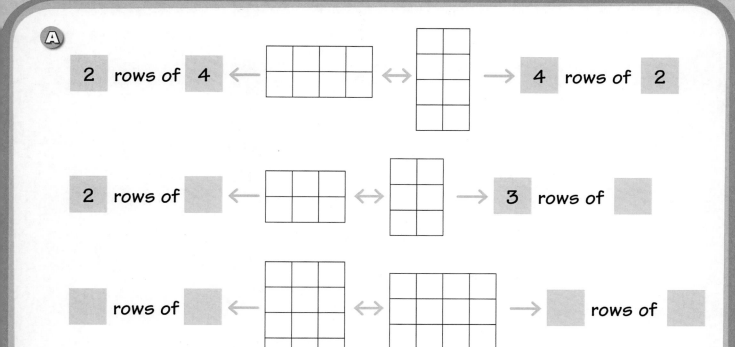

2 rows of 4 ← ↔ → 4 rows of 2

2 rows of [ ] ← ↔ → 3 rows of [ ]

[ ] rows of [ ] ← ↔ → [ ] rows of [ ]

B John has 9 bags of oranges and each bag has 10 oranges in it. Jane has 10 bags of oranges and each bag has 9 oranges in it. Who has more oranges altogether, or do they both have the same amount?

Answer:

C

3 × 10 = 30 ⟶ 10 × 3 =

6 × 5 = ⟶ 5 × 6 =

7 × [ ] = 70 ⟶ 10 × [ ] =

D One hundred children each uses a five dollar note to pay for a trip to the zoo. In the school office, the secretary starts to count the money by skip counting by fives. Then she realises there is a quick way to find the total. What is the total money collected?

Equation: 100 × [ ] = [ ]

Introducing fractions

I find half of a number by colouring in equal sets.

A

 Half of 10 =

Half of ☐ = ☐

Half of 14 = ☐

Half of ☐ = ☐

Half of ☐ = ☐

B

Half of 18 = ☐   Half of 10 = ☐

Half of 14 = ☐   Half of 20 = ☐

C Marnie has $40. She spends half of the money on fruit. How much money does she have left?

Answer: $ ☐

Fraction of a region or a set

I know most fraction words end in *th*: fourth (or a quarter), fifth, sixth, seventh, eighth, ninth, tenth ...

3 eighths	← shaded		unshaded →	eighths
	← shaded		unshaded →	tenths
	← shaded		unshaded →	fifths
	← shaded		unshaded →	
	← shaded		unshaded →	quarter
	← shaded		unshaded →	
half	← shaded		unshaded →	

The fraction words and their symbols

I always read a fraction like $\frac{5}{9}$ as five ninths.

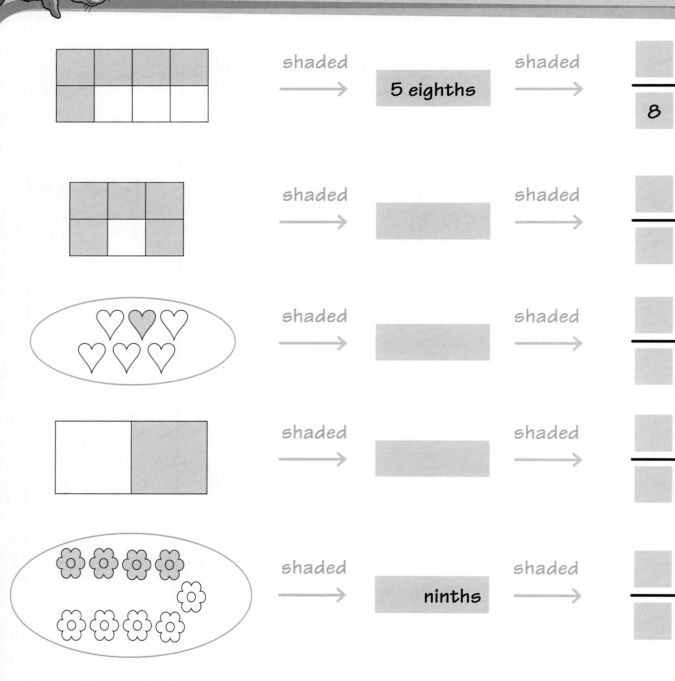

shaded → 5 eighths → shaded → $\dfrac{}{8}$

shaded → → shaded →

shaded → → shaded →

shaded → → shaded →

shaded → ninths → shaded →

shaded ← ← shaded ← $\dfrac{3}{4}$

Adding and subtracting fractions with the same denominator

I add and subract fractions by using the basic facts.

A

1 quarter + 2 quarters = $\dfrac{}{}$ ⟶ $\dfrac{1}{4}$ + $\dfrac{2}{4}$ = $\dfrac{}{}$

1 seventh + 4 sevenths = $\dfrac{}{}$ ⟶ $\dfrac{}{7}$ + $\dfrac{}{7}$ = $\dfrac{}{}$

6 ninths – 2 ninths = $\dfrac{}{}$ ⟶ $\dfrac{}{9}$ – $\dfrac{2}{}$ = $\dfrac{}{}$

B Write a story for 3 sixths + 2 sixths.

Write your question:

Write the answer:

Write the equation: $\dfrac{}{6}$ + $\dfrac{}{6}$ = $\dfrac{}{}$

Skip counting forwards and backwards by fives

5	25	45		85	105			165	
10	30		70						
15			75						
20	40	60					160	180	200

105	110	115	120						
155	160								
205				225					
255	260				280				300

300	295	290	285			270			
250		240		230					
150	145			130				110	

Four timed tests: one minute per test, one test per day

A

+	2	7	4	9
8				
6				
4				
7				
2				
5				

Correct: ____ out of 24

B

+	3	8	5	1
9				
7				
5				
8				
3				
6				

Correct: ____ out of 24

C

+	4	9	6	2
8				
6				
4				
9				
5				
7				

Correct: ____ out of 24

D

+	5	1	7	3
8				
6				
4				
7				
2				
5				

Correct: ____ out of 24

Speed test graph

After each one-minute test, mark the number you got correct with a line. Colour in the bar below your line. Then write the date.

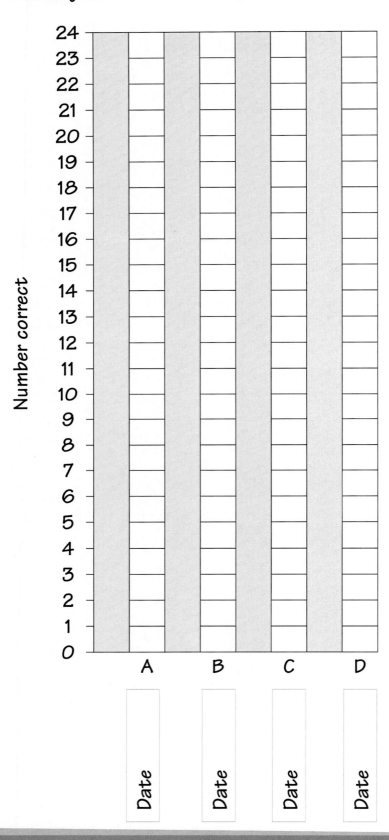